Muscle Contraction

SECOND EDITION

Clive R. Bagshaw
Reader in Biochemistry
University of Leicester

Published by Chapman & Hall, 2–6 Boundary Row, London SE1 8HN

Chapman & Hall, 2–6 Boundary Row, London SE1 8HN, UK

Blackie Academic & Professional, Wester Cleddens Road, Bishopbriggs, Glasgow G64 2NZ, UK

Chapman & Hall, 29 West 35th Street, New York NY10001, USA

Chapman & Hall Japan, Thomson Publishing Japan, Hirakawacho Nemoto Building, 6F, 1-7-11 Hirakawa-cho, Chiyoda-ku, Tokyo 102, Japan

Chapman & Hall Australia, Thomas Nelson Australia, 102 Dodds Street, South Melbourne, Victoria 3205, Australia

Chapman & Hall India, R. Seshadri, 32 Second Main Road, CIT East, Madras 600 035, India

First edition 1982
Second edition 1993

Typeset in Times 11/13pt by ROM Data Corporation Ltd, Falmouth Cornwall

Printed in Great Britain at the University Press, Cambridge

ISBN 0 412 40370 6

A catalogue record for this book is available from the British Library

Library of Congress Cataloging-in-Publication data available

∞ Printed on permanent acid-free text paper, manufactured in accordance with the proposed ANSI/NISO Z 39.48-199X and ANSI Z 39.48-1984

To Dana

Contents

Preface to the second edition

As one of the Outline Studies in Biology series, the first edition of *Muscle Contraction* was restricted to 80 pages. Although this demanded a considerable degree of editing, far more than any author would have liked at the time, it did bring the benefit of focusing my attention on what seemed to be the most important aspects of a very broad subject. To the reader the chapters became manageable in one sitting. The current edition did not have such a precise page limit, but I have borne in mind that the most favourable comments received concerning the first edition made some reference to the book's shortness. Nonetheless, the additional space has allowed me to introduce some of the advances of the last decade, including the development of *in vitro* motility assays and molecular genetic approaches. At the same time these novel techniques have brought a new readership from other fields of biology and therefore I have retained much of the introductory material regarding the more classical approaches.

The first edition was conceived when I first moved to Leicester and was awaiting delivery of equipment for my laboratory. The second edition was written in much less frustrating times. I am grateful to Professor Jim Spudich for allowing me to join his laboratory whilst on sabbatical. He and other members of the Department of Cell Biology at Stanford provided a stimulating environment for discussions and an opportunity to learn about some of the most recent developments first-hand. In particular, I am indebted to Dr Hans Warrick, Dr Taro Uyeda,

Kathy Ruppel, Ken Niebling and Jeff Finer for their help and comments on the first draft. I am also grateful for discussions with and comments from Dr Neil Millar, Professor Bob Simmons, Dr Roger Cooke, Dr Toshio Yanagida, Dr John Kendrick-Jones, Dr Rob Cross, Dr Ian Trayer, Dr John Sparrow, Dr Michael Geeves, Dr Bernhard Brenner and Dr Peter Knight. The task of illustrating the book was made much easier by the photographs and diagrams kindly provided by Professor Ken Holmes, Dr Ron Milligan, Professor Basil Northover, Dr Hans Warrick, Dr John Squire, Dr Jeff Harford, Dr Mary Reedy, Dr Avril Somlyo, Dr Darl Swartz, Dr Marion Greaser, Dr Peter Knight, Dr Gerald Offer, Dr Roger Craig, Dr Peter Vibert, Dr John Kendrick-Jones, Dr Andrew Jackson, Dr Don Winkelmann, Dr Andrew Sowerby and Dr Richard Ankrett. I am grateful to the Science and Engineering Research Council for funding my travel to Stanford University.

Permission to reproduce copyrighted material from the following publishers is gratefully acknowledged. The Physiological Society (Figs 2.6, 6.1, 6.11, 6.12, 7.6), The Rockefeller University Press (Figs 3.5, 4.8, 8.3), Academic Press (Figs 3.4, 4.4, 4.15, 9.1, 9.2), Macmillan Press (Figs 4.2, 4.3, 4.12, 4.13, 6.6, 6.9, 7.8, 9.3), Longman Group (Fig. 6.7), The Royal Society (Fig. 3.7) and D. W. Fawcett (Fig. 3.1).

Preface to the first edition

The topic of muscle contraction attracts interest from many disciplines – physiology, biochemistry and biophysics among them. Dialogue between these fields has always been strong but in formulating a molecular mechanism of contraction they have fused. However, the literature has tended to remain scattered between journals and polarized between the level found in general textbooks and that in specialist reviews. My object in writing this book is to provide a bridge to link these sources. No doubt you have already glanced at the contents and have discovered my prejudices. I hope to have struck an acceptable and useful balance, but as a precaution the references given as further reading either extend my treatment or fill in on background knowledge that I have assumed.

I am indebted to Sir Andrew Huxley, Dr Richard Tregear, Dr David White, Dr Arthur Rowe, Dr Arthur Moir, Dr Michael Geeves, Dr Christine Wells, Kate Poole and Simon Byrne for their comments on one or more sections of the first draft.

Introduction

1

The ability to move is one of the fundamental characteristics of a living organism, but the mechanisms by which this is achieved are diverse. The most familiar system of higher organisms involves the contraction of muscle and, indeed, study of this specialized tissue has opened up our understanding of more basic mechanisms of cell motility. This situation contrasts with many other areas of biology where the most detailed information, at the molecular level, was first obtained from the study of the most primitive organisms.

It is constructive to consider why muscle, the striated variety in particular, has led to this insight and remains an appealing system for investigation. Firstly, a large proportion of the cell material is devoted to contractile function. The two fundamental proteins involved, actin and myosin, comprise up to 80% of the structural proteins present and are therefore available in large amounts for biochemical characterization. Secondly, these proteins are arranged in a regular way so as to provide clues about the way in which their interaction produces force and motion. Furthermore these regularities have allowed the application of diffraction techniques to living muscle which complement and extend the structural information obtained by microscopy. Thirdly, contraction occurs on a macroscopic scale. In particular, the unidirectional contraction of a skeletal muscle along its long axis facilitates the quantitative measurement of length and tension, which can be related to events at the molecular level.

Nevertheless, the next decade is poised to challenge these favourable aspects, as new techniques are developed which tackle the problem of energy transduction directly at the molecular level. Will the mechanistic basis of muscle contraction be solved from the study of non-muscle systems? Recent developments outside the field are clearly going to influence the advancement and testing of models for contraction. However, the long history associated with the study of muscle has led to a framework of ideas which should not be overlooked, although all are open to question. There is good reason to believe developments in muscle and motility will go hand-in-hand as the new technologies unfold.

The focus of this book is the molecular basis of the contraction and regulation of the myofibril – the organelle within a muscle cell that is responsible for movement. Equally important problems from the point of view of understanding the overall performance of a muscle are the mechanisms of nervous and hormonal control, the back-up metabolic processes which supply the fuel for contraction and the gross anatomical features which lead to the movement of the tissue or organism as a whole. These topics are not covered in detail but further introductory literature is referenced in Chapter 2.

Interest in the mechanism of contraction at the molecular level extends beyond that of muscle action *per se* in several directions. Actin and myosin are ubiquitous within eukaryotic cells where they may be involved in cell locomotion and shape changes, or the movement of organelles within them. Indeed a striated muscle may be viewed as being at one end of a spectrum in which the myofibrils are relatively stable structures, whereas in non-specialized cells the contractile apparatus is assembled and disassembled as required. Furthermore, it is likely that the basic mechanism of mechanochemical coupling between actin and myosin also applies to other so-called motor proteins. Dynein and kinesin are involved in translocation events along microtubules and provide a complementary system for driving other forms of cell motility and organelle transport. At the chemical level, parallels are also observed between actomyosin and other adenosine 5′-triphosphate (ATP) utilizing enzymes involved in energy transduction, in particular those which give rise to or are driven by concentration gradients. A common feature found here is the ready reversibility of the ATP hydrolysis step at the active site, whilst the ATP binding step is often coupled to associated endergonic reactions.

The components of striated muscle have long provided a challenging system in which to investigate the principles of biological assembly. More recently it has become a favoured system for the study of gene switching as the tissue differentiates and matures to the adult form. Each protein isoform which is sequentially expressed is tuned to the physical and chemical performance required of it. However, the signals which trigger this differentiation are poorly understood. The extreme specialization of the mature muscle cell is a reflection of the demands placed on it, not only in generating force but in the rapid switching between active and relaxed states. The metabolic activity of a muscle can change by several orders of magnitude in a fraction of a second. Here the signal has been shown to involve a change in Ca^{2+} concentration. Indeed, the importance of the Ca^{2+} ion as a chemical messenger was first identified in muscle, and led the way to the discovery of a whole family of Ca^{2+}-binding proteins whose distribution is ubiquitous. These proteins are responsible not only for the activation of contractile proteins, but they also coordinate the activation of metabolic enzymes involved in providing the fuel supply. In other cells, related Ca^{2+}-modulated proteins regulate secretion and division. It is likely that the allosteric mechanisms involved in the transfer of information from a Ca^{2+}-binding site to a neighbouring protein share common features in all these systems.

It is clear that the challenge and diversity of the problems posed in unravelling the mechanism of muscle contraction will ensure the field remains active and multidisciplinary for the foreseeable future.

TOPICS FOR FURTHER READING

Amos, L.A. and Amos, W.B. (1991) *Molecules of the Cytoskeleton*, Macmillan, Basingstoke and London. (A succinct introduction to contractile proteins in the context of cells in general.)

Goldspink, G. (1992) The brains behind the brawn. *New Scientist*, **1832**, 28–33. (A readable account of the adaptations of muscles to the physical demands placed on them.)

2 Gross anatomy and physiology

2.1 MUSCLE TYPES

Muscles have evolved to meet a wide variety of functions that demand gross differences in performance. Skeletal muscles may be required for short bursts of activity or prolonged contractions. Sustained activity is the hallmark of cardiac muscle which can function continuously for more than a century. Some insect flight muscles are adapted for speed and can contract 1000 times a second, while molluscan adductor muscles are built for strength and can lift a 10 kg weight per square centimetre cross-section.

Both the common and specialized properties of these muscles are of interest. The former lead us to the essential elements of the contractile mechanism, while the latter may make particular features amenable to experimental investigation. All muscles involve the interaction between actin- and myosin-containing filaments fuelled by ATP hydrolysis. However, the filaments may differ in their arrangement and in the protein isoforms they contain. Muscles also differ in the metabolic reactions that they employ to generate ATP and in the way that they are controlled by, or respond to, nerve impulses and chemical effectors. Muscles may be classified by any one of these characteristics and this results in an overlapping and confusing array of nomenclature.

Vertebrate striated muscles take on their striped appearance under the light microscope because of the alignment of their myofilaments, whereas vertebrate smooth (unstriated) muscle cells appear almost structureless at this resolution. The latter class comprise a diverse collection of tissues which are generally under involuntary control and are concerned with the slow contraction and constriction of internal organs and vessels. Striated muscle may be further divided into skeletal and cardiac forms which differ in their anatomical arrangement and their excitation mechanism. The former is under conscious (voluntary) control, whereas the latter undergoes regular, self-sustained contractions. Individual fibres of a skeletal muscle may respond in an 'all-or-none' fashion to a nervous impulse (twitch) or they may be multi-innervated and produce a graded (tonic) response. However, a muscle comprising twitch fibres can yield a graded response by changing the number of individual fibres which are activated. The time course of tension development by a twitch fibre may be classified as fast or slow [see Wilkie, 1976; Aidley, 1989].

The colour of a muscle also provides means for its functional classification. Red muscle has a high content of cytochromes and myoglobin, proteins which are associated with oxidative metabolism, and allow sustained activity. White muscle relies on glycolysis for rapid ATP synthesis and readily goes into oxygen debt. It is therefore associated with fast-twitch fibres which operate in short bursts [see McMahon, 1984]. On cooking, these muscle types give rise to the familiar grey and white meat respectively. Many muscles contain a mixture of these fibre types and thus are not suitable for some experimental investigations.

The above classifications refer to vertebrate muscles, although invertebrate muscles show analogous characteristics [see Aidley, 1989]. They also show some apparently novel ones. Some insect flight muscles oscillate far more rapidly than the frequency of the nervous impulse which stimulates them, hence the coupling is termed asynchronous. The muscle itself only shortens by a few percent during each oscillation but the movement is amplified by a lever system involving the thorax. The wings and the attached ligaments are driven at their natural frequency of oscillation and hence emit a monotonous buzz. The nature and arrangement of the filaments of some invertebrate muscles have evolved to achieve great strength. Their myosin filaments are thicker and may be bolstered with a core protein, paramyosin. Molluscan smooth muscles have developed a mechanism for holding very high tensions with little energy expenditure, the so-called 'catch' state. More recently a somewhat similar property has been characterized in vertebrate smooth muscle and is termed the 'latch' state.

For the cases analysed so far, each muscle type from each species has one or more characteristic myosin isoforms, differing in amino acid composition and ATPase activity. Other contractile proteins may show similar diversity in sequence although actin is relatively well conserved. Ultimately sequence information should allow a molecular classification of muscles, but at present many of these isoforms have only been resolved electrophoretically and are named according to their source using the terminology outlined above.

It is pertinent to consider some specific muscles which are widely studied at the molecular level [see Squire, 1986, Chapter 3]. In Chapter 1 the advantages of striated muscle for experimental investigations were outlined. The frog sartorius is favoured for many physiological and structural studies. It is small enough to allow adequate oxygenation and contains fuel reserves for many hundreds of contractions. For single fibre work, the semitendinosus muscle is preferred because it is more readily dissected and can be stretched to prevent filament overlap, a condition which abolishes actomyosin interaction. Frog muscles are not favoured for biochemical studies, because of the limited quantity and instability of the isolated proteins. For these experiments the usual source is rabbit back (longissimus dorsi) and some leg muscles which comprise largely fast-twitch fibres. Intact mammalian muscles are not as easy to dissect and they do not survive as well as frog muscle. Nevertheless, rabbit psoas muscle is often favoured as it is relatively free of connective tissue and has parallel running fibres. It has therefore been widely used for physiological and structural studies. Throughout this book (and most other introductory literature) whenever the muscle source is undefined, it should be regarded as being skeletal muscle from one of these species. Generalization of the data to other types may not always be valid. Comparisons between rabbit and frog muscles show some quantitative differences, but these are usually within the variations arising from different experimental conditions. In this context many of the numbers given in this book are approximate and are used primarily to exemplify the nature of a calculation rather than to report a specific experimental result.

Cardiac and smooth muscles are not so conveniently investigated, but because of their medical importance they have received enormous attention. Bovine hearts provide ample biochemical material, while the papillary muscles within the hearts of smaller mammals are often the choice for physiological studies because of their ease of isolation. The choice of smooth muscle is normally dictated by the question under investigation, but frequently used sources are the walls of the gut and blood vessels. For large scale biochemical characterization, chicken gizzard is a favoured source, while single smooth muscle cells may be isolated from a toad stomach by enzymic digestion.

Invertebrate muscles, besides their inherent interest, provide some practical benefits [see Rüegg, 1986, Chapter 8]. Insect flight muscle is comparatively well preserved during the preparation for electron microscopy and its study has led to some important concepts in the molecular mechanism of contraction. The giant water bug, *Lethocerus*, is a useful source. Certain crustacean muscles (e.g. crab and giant barnacle) contain exceptionally large muscle fibres and long sarcomere spacings which aid microinjection techniques. These were instrumental in establishing the role of Ca^{2+} in activation of contraction. Molluscan muscles have been investigated in an attempt to elucidate the catch mechanism–the anterior byssus retractor muscle of the mussel being of suitable dimensions for physiological and structural work. Molluscan muscles also show a novel control mechanism that may a have wider distribution, and is particularly amenable to investigation in the scallop striated adductor muscle, owing to the ready and reversible dissociation of the regulatory subunits.

Genetic approaches are poised to make a major contribution to the elucidation of the contractile mechanism and already have provided unique insights into the assembly of sarcomeric structures. Mutants of the nematode worm, *Caenorhabditis,* led to identification and sequence of the first myosin gene. The 1980s witnessed an explosion in protein sequence information predicted from DNA sequencing, although traditional methods are still necessary to pinpoint sites of post-translational modifications. Detailed physiological studies have been made in conjunction with site-directed mutagenesis of contractile proteins of the flight muscles of the fruit fly, *Drosophila*, in an attempt to identify individuals with impaired muscle mechanics. One of the goals of this work is to perturb the actomyosin interaction so as to build up intermediate states which are present in vanishingly small concentrations under normal conditions. As techniques are developed to quantitate force and movement without the need for the sarcomeric arrangement of striated muscle, so more primitive sources of actomyosin are likely to come to the fore. The strong background of genetic manipulations in organisms such as the yeast, *Saccharomyces*, and slime mould, *Dictyostelium*, have led to new insights into contractile protein function in non-muscle cells [see Gerisch *et al.*, 1991, Chapter 10]. History has shown that the choice of the most amenable tissue or organism can often outweigh many years of technical development.

2.2 PHYSIOLOGICAL STATES

Relaxed muscle is readily extensible. In this condition the actin and myosin filaments do not interact strongly and the elasticity is provided by the other cytoskeletal proteins, the associated connective tissue and membranes. On stimulation, the response of the muscle depends on the external constraints. If the muscle is held at a fixed length it will develop tension. Such a process is termed an isometric contraction, even though the muscle does not actually contract (i.e. shorten) in the familiar sense. If the load attached to a muscle is less than the isometric tension, the muscle will shorten. The steady velocity of such an isotonic (i.e. constant load) contraction reaches a maximum (V_o) with zero external load. Applying a force greater than the isometric tension (P_o) will cause the muscle to extend. Its resistance becomes rather low when the applied load is greater than $2P_o$. A muscle *in vivo* may experience all of these conditions during its normal functioning. Muscles can only actively shorten. *In situ*, they return to their resting length under the influence of an external force provided by an antagonistic muscle or some elastic structure. Muscles increase their girth on shortening to maintain a constant volume [see Wilkie, 1976].

If metabolic events are curtailed so that the ATP is not replenished, a muscle becomes stiff – a state known as rigor. Attempts to stretch a rigor muscle quickly by more than a few percent of its length results in permanent damage to the fibres. Although rigor is a non-physiological condition, it has received a great deal of attention because it is a stable (and hence experimentally amenable) state in which the actin and myosin filaments interact strongly. If the outer membrane of the muscle cell is rendered permeable, addition of ATP will cause a rigor muscle to revert to a functional state. It will either contract or relax depending on the Ca^{2+} ion concentration.

2.3 ACTIVATION

Skeletal muscle is under nervous control and is electrically excitable. Let us consider the nature of the signal it receives. The cytoplasm of resting nerve and muscle cells is at a potential of -60 to -90 mV with respect to the bathing medium. This voltage exists because certain ions are prevented from equilibrating across the membrane barrier. Freely permeable ions take up a Nernstian distribution, which represents a balance between their electrical and osmotic energies. In a resting nerve, the ratio of external to internal K^+ concentration is about 0.1 and suggests that this ion is almost at equilibrium. In contrast, Na^+ is maintained at a ratio of 10 by an active sodium pump. On excitation, the membrane of a nerve cell becomes locally and temporarily permeable to Na^+, which rapidly enters down its concentration gradient, causing the membrane potential to rise to + 40 mV. However, within a few milliseconds K^+ outflow restores the potential to its resting value. This transient depolarization excites the neighbouring region of the nerve cell, resulting in self-propagation of the signal (the action potential). The amount of Na^+ and K^+ actually exchanged, relative to the total contents of these ions, is minute because the electrical capacity of a cell is very low [see Aidley, 1989].

The nerve terminates in a motor end plate which abuts against the muscle cell. Here the action potential causes the release of a transmitter substance, acetylcholine, which in turn depolarizes the muscle cell membrane. In the case of a twitch fibre, the action potential is propagated along its length as described for the nerve cell. The fibre membrane is highly invaginated so that the electrical signal is carried inwards and ultimately stimulates the adjacent vesicles of the sarcoplasmic reticulum to release Ca^{2+} (section 3.3). The free Ca^{2+} concentration in the cytoplasm (sarcoplasm) of the muscle cell rises from about 0.1 μM to 10 μM. Relaxation is achieved by active reaccumulation of the Ca^{2+} by the sarcoplasmic reticulum. Other muscle types may be activated by different mechanisms. In some smooth muscles the membrane is insensitive to electrical events, but may respond to specific activators such as adrenaline. Nevertheless the ultimate message received by the myofilament proteins is usually a rise in $[Ca^{2+}]$ and this finding allows the electrical events of the excitation–contraction coupling mechanism to be considered as a separate problem [see Rüegg, 1986, Chapter 8].

An isolated twitch muscle may be stimulated via its attached nerve or, more directly, by applying a small electric shock across its membrane.

A characteristic of twitch fibres is that they are maximally activated above a certain threshold stimulus strength – an 'all-or-none' response. In order to activate all the fibres within the muscle it is necessary to apply a supramaximal shock. A single shock causes a twitch contraction. Repeated shocks cause multiple twitches and above a characteristic stimulus frequency these fuse to give a tetanic contraction which remains fairly steady until the stimuli are withheld or until fatigue sets in (Figure 2.1). Frog fast skeletal muscle shows an initial lag in tension development of several milliseconds due to the activation mechanism and develops a peak tension in about 50 ms [see Wilkie, 1976].

If the membrane systems of a muscle cell are damaged, the muscle may no longer be electrically excitable. Instead, the concentrations of Ca^{2+}, Mg^{2+} and ATP in the vicinity of the myofibrils can be controlled directly by the composition of the external bathing medium. Figure 2.2 shows the relationship between the state of a demembranated muscle and the composition of the medium. In general, muscle control systems are so sensitive to Ca^{2+} that contaminating amounts in buffers and glassware are sufficient for activation. The Ca^{2+} chelator, ethyleneglycol-bis-(β-aminoethyl ether)- tetraacetic acid (EGTA) is routinely added to reduce the free $[Ca^{2+}]$ to < 1 μM in order to achieve the relaxed state. Unlike ethylene diamine tetraacetic acid (EDTA), EGTA does not bind Mg^{2+} with high affinity. As discussed later, a high free $[Mg^{2+}]$ is required for

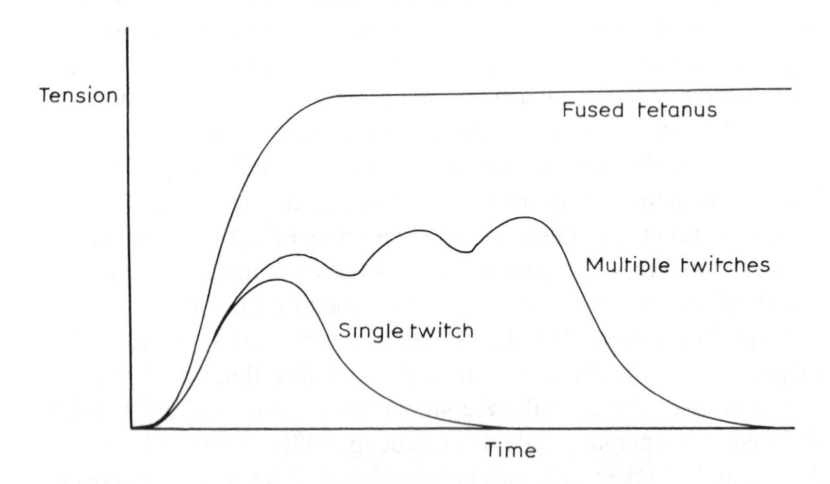

Fig. 2.1 The activation of skeletal muscle by single and repetitive stimulation. A fused tetanus occurs at a stimulus frequency greater than about 10 s^{-1}

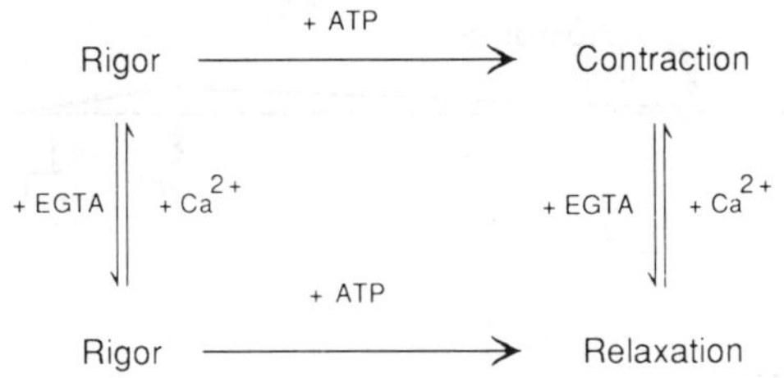

Rigor —+ ATP→ Contraction

+ EGTA ⇅ + Ca^{2+} + EGTA ⇅ + Ca^{2+}

Rigor —+ ATP→ Relaxation

Fig. 2.2 The physiological states of a demembranated muscle fibre.

both contraction and relaxation because the actual substrate for myosin is Mg.ATP.

The site of action of Ca^{2+} varies in different muscles. The most well characterized system involves the regulatory protein, troponin, which intrinsically suppresses the interaction between the actin and myosin filaments. Ca^{2+} acts, in effect, by rendering the troponin system inoperative. Removal of this regulatory protein therefore results in permanent activation regardless of the Ca^{2+} concentration. This is a useful characteristic because it allows purified mixtures of actin and myosin to be studied, as a model for contraction, independently of the mechanism of regulation. However, not all regulatory systems function in this way [see Rüegg, 1986, Chapter 8].

2.4 PHYSIOLOGICAL PERFORMANCE

A quantitative assessment of the physiological states alluded to in section 2.2 remains central to the problem of muscle contraction. A knowledge of muscle performance both defines the problem and provides some clues to feasible answers. Apparatus for measuring length and tension has developed to a high degree of sophistication with improved sensitivity, stability and response time. A simple mechanical device where a muscle operates against a weight illustrates the principles and problems (Figure 2.3). Such apparatus can be used to define the steady-state contractile properties of a muscle but its inertia does not allow much scope for transient analysis. Inertia may be reduced and sensitivity increased by the use of electronic and optical devices (section 6.5).

For many muscles the steady velocity of contraction depends on the load as shown in Figure 2.4. The velocity is at a maximum (V_o) when the

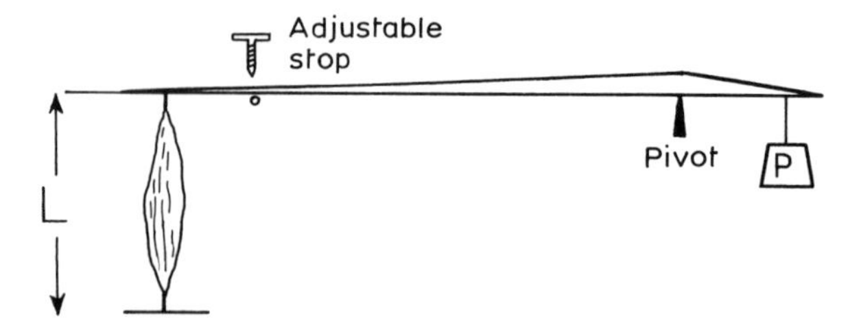

Fig. 2.3 A simple mechanical device for recording the length and tension of a muscle.

muscle is unloaded and declines to zero when the load matches the isometric tension (P_o). Note that a muscle only performs work when its tension, P operates over a distance. The power output ($P \times$V) reaches a maximum when P and V are about one third their maximum values – a fact that cyclists should be aware of when selecting the best gear. When exerting its isometric tension (P_o) or contracting at its maximum velocity

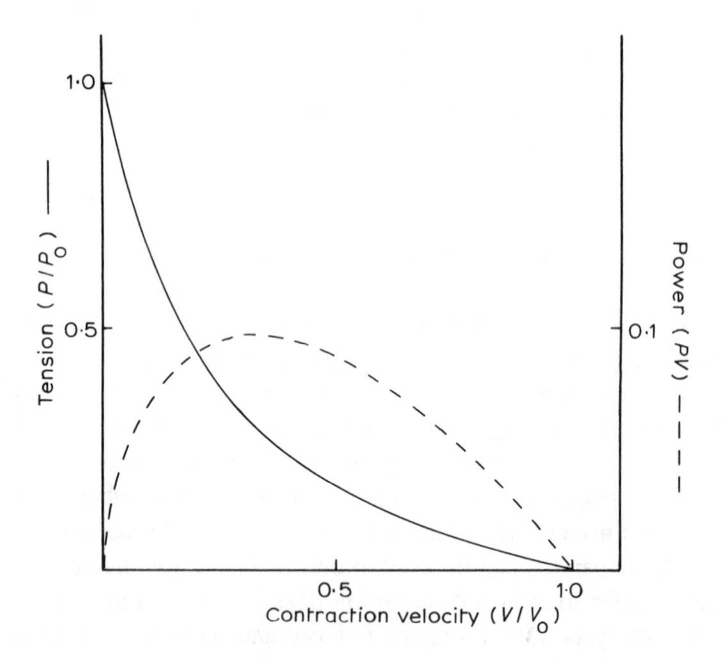

Fig. 2.4 Force (tension)–velocity curve of skeletal muscle. Stretching the muscle (equivalent to a negative velocity of contraction) causes a sharp rise in tension up to about $2P_0$ before it yields.

Table 2.1 Mechanical properties

Term	*Symbol*	*Definition*	*SI units*	*Comments*
Length	L		m	Expressed relative to rest length or as a sarcomere length.
Tension (Force)	T	mass × acceleration	$N = kg\ m\ s^{-2}$	T = internal force which resists applied load (1 kg weight = 9.8 N).
	P	tension/unit area	$N\ m^{-2}$	Normalized for cross-sectional area.
Stiffness	S	T/L P/L	$N\ m^{-1}$ $N\ m^{-3}$	Dimensions depend on whether length and tension are normalized.
Elastic modulus	E	P/unit length	$N\ m^{-2}$	
Compliance	C	L/T	$m\ N^{-1}$	Reciprocal of stiffness.
Velocity	V	length change/time	$m\ s^{-1}$	May be normalized to muscle length or refer to half-sarcomere.
Work	W	force × distance	$J = N\ m$	May be normalized to sample of unit dimensions.
Power		work/time	$W = J\ s^{-1}$	For constant P and V, power = $P.V$
Efficiency		power/chemical energy flux		Chemical energy flux calculated from ATPase.
Economy		tension/chemical energy flux	$N\ s\ J^{-1}$	Reciprocal defined as tension cost

(V_o), a muscle does no external work (= a force acting over a distance) but it continues to utilize chemical energy and hence has zero efficiency. The cost of maintaining a steady tension is, however, an important parameter of a muscle which has evolved to bear loads and is defined as its economy. By stretching a muscle, the P–V relationship (Figure 2.4) has been extended to negative velocities. Under these conditions the force rises sharply until it equals about $2P_o$, then a yield point is reached when the resistance to stretch becomes low. Table 2.1 summarizes the definitions of mechanical parameters, while some illustrative data for frog sartorius muscle are given in the Appendix (Table A.1).

2.5 FUEL AND ENERGETICS

The energy source for contraction is derived from the reaction:

$$Mg.ATP^{2-} + H_2O \rightleftharpoons Mg.ADP^{1-} + Pi^{2-} + H^{+} \quad (2.1)$$

as demonstrated by the action of MgATP on muscle fibre preparations rendered permeable to the bathing medium. The reaction occurs at sites on the myosin filament and is activated by the actin filament. In intact muscle ATP utilization is not immediately apparent because of the extremely efficient buffering action of the creatine kinase reaction:

$$\text{Mg.ADP} + \text{phosphocreatine (PCr)} \overset{K\,=\,100}{\rightleftharpoons} \text{Mg.ATP} + \text{creatine (Cr)} \quad (2.2)$$

Several seconds after the initiation of contraction, glycolysis is activated and ATP is synthesized by substrate level phosphorylation [see Woledge *et al.*, 1985]. Phosphocreatine (PCr) is replenished during the recovery process by net reversal of equation 2.2. In some invertebrate muscles phosphoarginine takes on the equivalent role of phosphocreatine. ATP utilization can be demonstrated in contracting, intact muscle if creatine kinase is inhibited with fluorodinitrobenzene, glycolysis is blocked with iodoacetate and oxidative phosphorylation is prevented by using a nitrogen atmosphere. Under these conditions there is sufficient ATP to support eight or so twitches, whereas an uninhibited excised muscle has a glycogen store sufficient for several hundred contractions. White muscles rely on substrate level phosphorylation for the regeneration of ATP and can function anaerobically for brief periods at the expense of lactic acid production. They are the muscles used for sprints. Red muscles require a supply of oxygen to enable oxidative phosphorylation to proceed in the mitochondria. Long distance runners train to maximize blood flow to these muscles so that sustained activity is possible [see McMahon, 1984].

The ATP molecule (Figure 2.5) is a source of energy because the equilibrium constant for its hydrolysis, equation 2.1, lies far to the right. Several reasons for the increased stability of ADP and Pi (orthophosphate) relative to ATP have been advanced, including:

1. The resonance stabilization of the phosphate group. A crude estimate of this effect is obtained by comparing the number of equivalent

structures, for reactants and products, which may be written down using conventional single and double bond notation.

2. At physiological pH, cleavage of the β–γ bond of ATP relieves charge repulsion.
3. The extent of solvation increases on forming the product anions.

Note that the energetics depend on the phosphate moiety *in toto* and cannot be attributed to a specific high energy bond. The adenosine moiety does not contribute significantly to the energetics, but is important for specific recognition by the ATPase active site. The free energy available from hydrolysis depends on the degree to which the product–reactant concentration ratio, Q, differs from its equilibrium value:

$$\Delta G = RT \ln(Q/K) \tag{2.3}$$

Frequently the factor $RT\ln(1/K)$ is defined as the standard free energy change, ΔG_o, but care is required to avoid using this value out of context. ΔG_o is dependent on the dimensions of K and invites meaningless comparisons between reactions that differ in the number of reacting molecules (e.g. binding reactions cf. isomerizations).

The evaluation of ΔG for ATP in muscle therefore requires an appropriate *in vitro* determination of K and an *in vivo* determination of the concentrations of all the components involved (ATP, ADP, Pi, H^+, Mg^{2+} at a minimum). K is too large to be determined directly, but it can be calculated from the equilibrium constants of consecutive reactions which result in net ATP hydrolysis, e.g:

$$\text{ATP} + \text{glucose} \overset{K_1}{\rightleftharpoons} \text{ADP} + \text{glucose-6-phosphate} \overset{K_2}{\rightleftharpoons} \text{glucose} + \text{Pi} \tag{2.4}$$

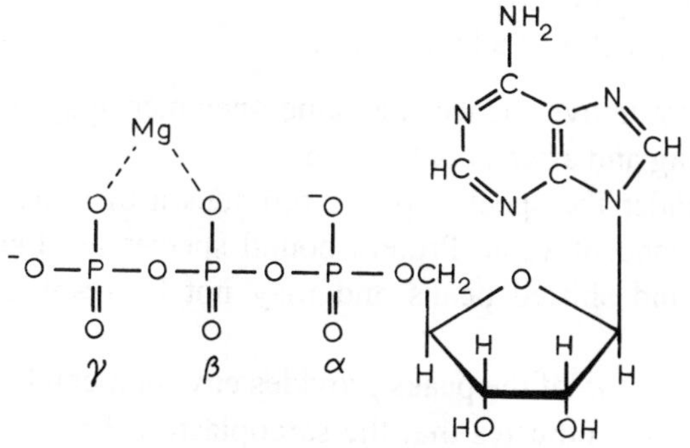

NH2
N
C
CH
HC
Mg
O
$^-$O
$^-$O — P — O — P — O — P — OCH2
O
γ
β
α
H
HO
OH

Fig. 2.5 The complex formed between Mg^{2+} and adenosine 5′- triphosphate (Mg. ATP).

hence, $K = K_1 . K_2$. The value of K depends on the Mg^{2+} and H^+ concentrations and under conditions appropriate to muscle [ADP].[Pi]/[ATP] = 10^6 M at equilibrium. In this definition [ADP] etc. includes all relevant species, i.e. $MgADP^-$, $HADP^{2-}$ etc. at the pH under consideration. H_2O is normally omitted from this calculation because its concentration remains practically constant (about 55 M) whether far from or at equilibrium.

The measurement of the actual concentrations of the relevant components in muscle is far from trivial and controversies remain over some of the values [see Woledge *et al.*, 1985]. Analysis in terms of mole per wet weight of tissue provides some information, but the effective free concentration in the vicinity of the myofibrils may be higher or lower than this value owing to compartmentation or binding. Such an analysis is carried out by clamping a muscle between metal tongs, which are precooled in liquid nitrogen, so that reactions are terminated as quickly as possible (within about 80 ms). The frozen tissue is pulverized and thawed in cold perchloric or trichloroacetic acid and the denatured protein is removed. The remaining metabolites are then analysed quantitatively using appropriate methods such as high performance liquid chromatography (HPLC). A large number of muscles are required for the construction of a time course of metabolite levels during contraction by this method. Complementary techniques are therefore important, in particular those which determine free rather than total concentrations.

Nuclear magnetic resonance (NMR) spectroscopy has been applied to this problem and much useful information has come from ^{31}P spectra of living muscle [1]. In a strong magnetic field, the ^{31}P nucleus (the natural isotope) absorbs radiowaves at a characteristic frequency dependent on its chemical environment. PCr, Pi and the three nuclei in ATP give well resolved peaks (Figure 2.6).

The advantages of the technique are:

1. It is non-destructive, hence the same specimen may be examined before, during and after a contraction.
2. The area under the spectral peaks provides a measure of the free metabolite concentration. Protein bound species are likely to have broadened and shifted peaks and may not be resolved from the background.
3. The exact position of the peaks provides environmental information, e.g. the Pi peak indicates that the sarcoplasm pH is 7.2 to 7.4 in a

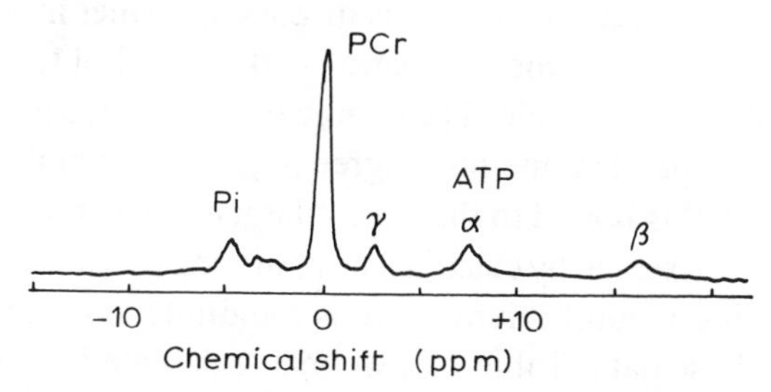

Fig. 2.6 ^{31}P NMR spectrum of frog sartorius muscle. (From Dawson *et al.* [1].)

non-fatigued muscle and the ATP peaks suggest this component is predominantly complexed with Mg^{2+}.

4. It may provide information about compartmentation if, for example, Pi is present in pools of differing pH.
5. Fluxes between metabolites can sometimes be evaluated by selectively irradiating one peak and observing the effect on other peaks in the spectrum (saturation transfer). This is appropriate to fluxes on the same time scale of the magnetic recovery process (around 1 s) such as the creatine kinase reaction.

The main disadvantage of NMR is its lack of sensitivity. Several minutes are required to build up sufficient signal to detect components in the millimolar concentration range. However, a time resolution of about 1 s can be achieved by summing many repeated contractions. When a muscle fatigues, the PCr peak decreases and the Pi peak rises concomitantly and may shift in the acid direction. The ATP peaks remain almost unchanged in intensity because of the efficiency of the creatine kinase reaction (equation 2.2).

Intracellular pH can also be determined with a glass microelectrode. The measured voltage includes the membrane potential of the fibre (section 2.3), which must be recorded with a separate microelectrode and subtracted. Other ion-sensitive electrodes have been developed, including those sensitive to Mg^{2+} and Ca^{2+}. The concentration of these ions can also be estimated by injection of indicator dyes, with the advantage of good sensitivity and time resolution. The approach is, however, open to interference from other ions and perturbation of the spectra by protein

Table 2.2 Estimates of metabolite concentrations

	ATP	*ADP*	*Pi*	*PCr*	*Cr*	*Mg^{2+}*	*Ca^{2+}*
Total (mmol/kg)	5	0.8	3	25	13	10	1
Free (mM)	4	0.02	2	25	13	3	0.0001

binding. The use of Ca^{2+} indicators is discussed further in section 3.3.

Table 2.2 provides some estimates of the metabolite levels of the sarcoplasm of resting muscle. The degree of confidence is higher where a number of independent methods agree (e.g. ATP). On the other hand, much of the ADP is bound to the actin filament so that the free concentration has been assessed by calculation, assuming that the creatine kinase reaction remains at equilibrium. ADP is too dilute to be detected by ^{31}P NMR. From these data (Table 2.2) we may calculate from equation 2.3 that $\Delta G = -60$ kJ mol^{-1} for ATP hydrolysis within an unfatigued muscle. This represents the maximum work that can be derived, although in practice a muscle can only use a proportion of this energy.

The position of the equilibrium of a chemical reaction is determined both by the change in internal energy (enthalpy) and orderedness (entropy):

$$\Delta G = \Delta H - T\Delta S \qquad (2.5)$$

If ATP is hydrolysed in a calorimeter under conditions appropriate to a muscle, the internal energy will be liberated as heat from which ΔH is determined as - 48 kJ mol^{-1}. A hypothetical muscle working at 100% thermodynamic efficiency would therefore need to absorb 12 kJ of heat from the surroundings per mole of ATP hydrolysed [see Wilkie, 1976]. In practice muscles do not exceed about 50% efficiency, so that the free energy of a shortening fibre is dissipated partly as heat and partly as work. However, in all cases, the heat + work output must balance the change in the internal energy (ΔH) of the reactions involved, in accordance with the First Law of Thermodynamics. This relationship provides a check on our understanding of the reactions occurring in muscle. The heat + work output of a contracting muscle is largely accounted for by the extent of phosphocreatine breakdown, but the discrepancies are believed to be significant and indicate that other heat generating reactions are involved. In early literature, before the ΔG for ATP had been ascertained, the efficiency of muscle contraction was often defined in terms of (work output)/(work + heat output) which included an entropic term. Subsequently this definition was termed the 'mechanical efficiency' to distinguish it from the true thermodynamic efficiency [see Woledge *et al.*, 1985].

Even before phosphate metabolites were identified, measurements of heat changes demonstrated an important property of the energy transduction mechanism. The rate of heat liberation was found to be higher in a muscle undergoing shortening compared with one developing isometric

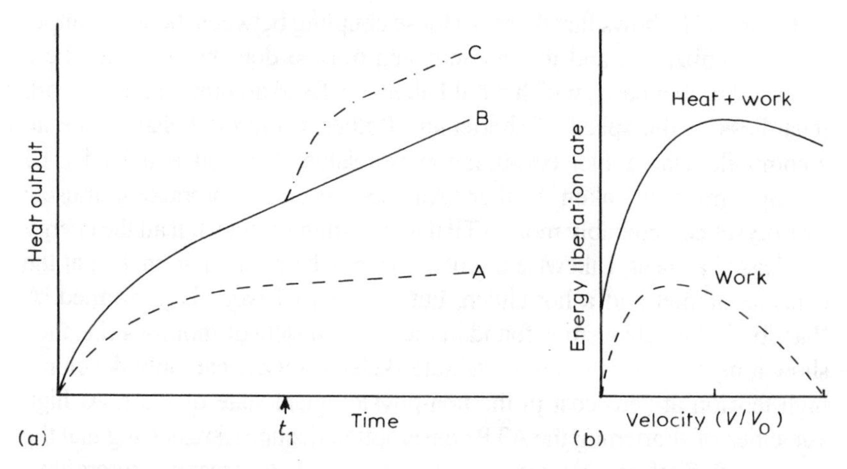

Fig. 2.7 (a) Heat output from a stimulated muscle (A) streched beyond filament overlap, (B) held isometrically and (C) allowed to shorten at time t_1. (b) Energy liberation rate as a function of contraction velocity.

tension – the so-called Fenn effect. Figure 2.7a depicts an idealized experiment of this type. Upon stimulation a burst of heat is liberated, associated with the activation mechanism, followed by steady heat produc- tion (curve B). The contribution of the activation heat may be assessed by stimulating a muscle which has been stretched so that the actin and myosin filaments no longer interact (curve A). If an isometrically contracting muscle is released at time t_1 and allowed to shorten, extra heat is released (curve C). Figure 2.7b shows the rate of total energy liberation (heat + work) and power output (work rate) as a function of the isotonic shortening velocity (cf. Figure 2.4). Freeze clamping experiments confirm that the heat + work output corresponds, to a first approximation, to the phosphocreatine used. The observed discrepancies may be related to changes in the populations of different mechanical states of the proteins under different contraction conditions. In particular, at high velocities of shortening, ATP consumption is relatively low although the heat output is high [see Woledge *et al.*, 1985]. The mechanisms involved in an unloaded shortening remain poorly understood (cf. section 10.1). From freeze clamping experiments the calculated ATP consumption by a frog sartorius muscle at 0°C undergoing isometric contraction is about 500 μmol s^{-1} kg^{-1} muscle, and may rise to three times this value on shortening at moderate velocities [2]. The ATP turnover by the contractile apparatus of resting muscle is very low and an upper limit of 0.5 μmol s^{-1} kg^{-1} may be calculated from the O_2 consumption [3].

Figure 2.7b shows that there is a loose coupling between the work output and fuel utilization and that an activated muscle does not operate like a stretched rubber band, which would liberate a fixed amount of heat + work regardless of the speed of shortening. Rather, a muscle behaves like an automobile where fuel consumption is related to speed and load. The analogy may be taken further. An isometrically contracting muscle hydrolyses considerably more ATP than a resting muscle, but all the energy is released as heat. Likewise a motor car may be held on an incline at the expense of fuel and a hot clutch, but no external work is performed. A 'handbrake' mechanism is found in the catch muscle of molluscs (i.e. they show a high economy), but vertebrate skeletal muscle can only develop a high tension at zero cost in the non-physiological state of rigor. At high velocities of shortening the ATP consumption declines, suggesting that the number of effective actin–myosin interactions falls. Car manufacturers have also recognized that cruising at high speeds does not demand as much fuel as during acceleration and have devised fuel injection systems which automatically reduce the number of cylinders which are fed.

The characteristic performance suggests and limits plausible mechanisms of contraction and should be borne in mind in the subsequent chapters which focus on the components involved. The procedures for preparing these components for biochemical investigations might be compared with stripping an engine with a sledge hammer. Even if the vital parts remain intact, we still need a guide as to how they may be assembled in a functional way.

TOPICS FOR FURTHER READING

Wilkie, D.R. (1976) *Muscle*, 2nd edn, Edward Arnold, London. (An introduction to muscle physiology.)

Aidley, D.J. (1989) *The Physiology of Excitable Cells*, 3rd edn, Cambridge University Press, Cambridge. (Provides a good introduction to excitation–contraction coupling.)

Woledge, R.C., Curtin, N.A. and Homsher, E. (1985) *Energetic Aspects of Muscle Contraction*, Academic Press, London. (A detailed review of chemical physiology.)

McMahon, T.A. (1984) *Muscle, Reflexes, and Locomotion*, Princetown University Press, Princetown, New Jersey. (A lucid account of muscle mechanics and what it means for the whole organism.)

Muscle cells

3

3.1 STRIATED MUSCLE

A schematic dissection of a vertebrate skeletal muscle is shown in Figure 3.1. At low power magnification, muscle is seen to comprise bundles of fibres sheathed in connective tissue. Each fibre is a giant multinucleated cell, formed by the fusion of many myoblast cells during its development. Individual fibres vary greatly in size from a few millimetres to many centimetres in length and from 20 to 100 μm in diameter. In some cases fibres may span the entire length of the muscle. Within each fibre are a thousand or so myofibrils, the rod-like organelles responsible for contraction which occupy up to 80% of its volume. Mitochondria and glycogen granules are sandwiched between the myofibrils [see Bloom and Fawcett, 1975].

A striking feature of each fibre is the presence of cross bands arising from aligned striations on each myofibril. Although myofibrils are transparent, these bands were seen by early microscopists by slightly defocusing, and they attributed this effect to regions of high refractive index (i.e. high protein concentration). Resolution of these bands by refraction rather than by staining is important because it allows living muscle to be observed. A number of microscopes have been developed for this purpose. Light passing through a region of high refractive index is slowed down and hence its phase is shifted relative to an uninterrupted beam. Phase contrast microscopy detects such a region, in thin specimens, by accentuating the interference between the diffracted light which passes through the region and the light which passes around its edge. With interference microscopy the illuminating beam is split by a half-

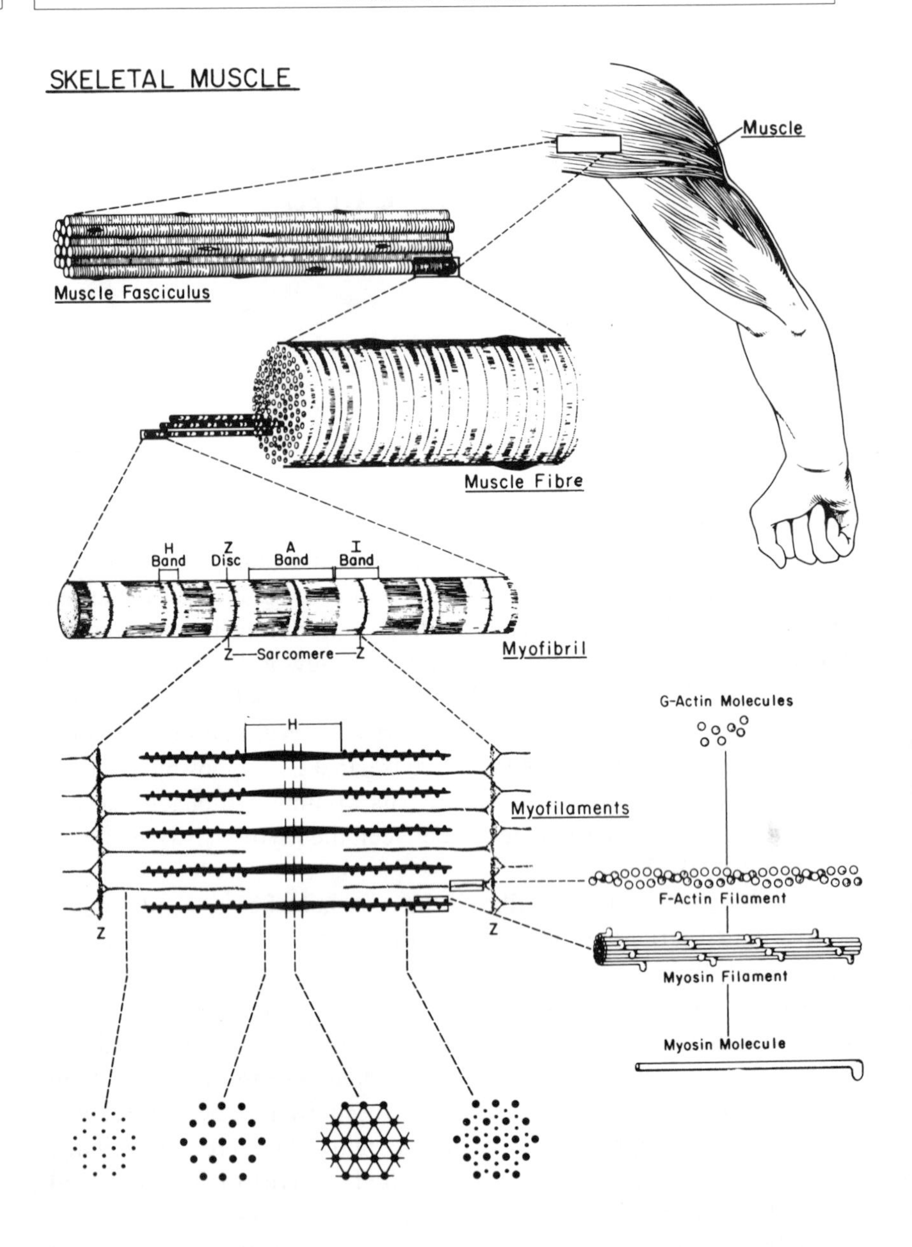

Fig. 3.1 The microanatomy of vertebrate skeletal muscle. [From Bloom and Fawcett (1975).]

silvered mirror and then recombined after passage of one beam through the specimen. This technique allows a quantitative estimate of the refractive index and may be applied to thicker specimens. If the elements (e.g. proteins) within a region responsible for its refraction are not distributed homogeneously, then the refractive index will depend on the plane of polarization of the light (birefringence). Such is the case with striated muscle. Under the polarizing microscope the protein dense bands are seen to be anisotropic with respect to refractive index, while the less dense regions between them are relatively isotropic, hence the terminology A- and I-bands. In the centre of the I-bands a highly refractive Z-line is observed, while the middle of the A-band is less dense, giving the H-zone.

The full significance of this banding pattern is revealed by the additional resolution of electron microscopy (Figure 3.2). Here it can be seen that the bands arise from the interdigitation of sets of filaments. Thin

Fig. 3.2 Electron micrograph of a longitudinal section of rabbit psoas muscle. (Courtesy of Dr R. Craig.)

filaments emanate from the Z-line and make up the I-band, while the thick filaments make up the A-band. The thin filaments extend into the A-band and so enhance its intensity, leaving the paler H-zone in the non-overlap region. In addition, an M-line is revealed in the centre of the H-zone which holds the thick filaments in register. The unit between two Z-lines is defined as a sarcomere. In cross-section the filaments are arranged hexagonally with one thick filament surrounded by six thin filaments. In non-vertebrate striated muscles the arrangement may differ. The distance between the filaments varies from about 20 to 30 nm depending on the sarcomere length.

When a muscle fibre is illuminated by a narrow beam of monochromatic light (e.g. a laser slide pointer), the striations give rise to an optical diffraction pattern. The distance between the fringes allows the calculation of the average sarcomere length. While this is a convenient, non-destructive method, care is required in critical studies because local areas of disorder may go undetected.

3.2 THE SLIDING FILAMENT THEORY

Using interference microscopy to view living muscle fibres, A.F. Huxley and Niedergerke [4] showed that, on stretching or shortening, the A-bands remain at constant length, while the I-bands change. Phase contrast microscopy by H.E. Huxley and Hanson [5] revealed that when isolated myofibrils were induced to contract by addition of ATP, the I-band and H-zone shortened in unison. These findings, taken in conjunction with electron micrographs which demonstrated the underlying basis for the striations, led to the sliding filament theory in which contraction was proposed to occur solely by interdigitation of thick and thin filaments. The sarcomere length depends on the degree of overlap while the filaments themselves remain at constant length. In frog skeletal muscle, the thick filaments, and hence the A-bands, are 1.6 μm long and the thin filaments extend 1.0 μm either side of the Z-line. At rest length, the sarcomere spacing is about 2.6 μm long and hence the overlap is 0.5 μm in each half-sarcomere.

The thick filament comprises mainly myosin and the thin filament mainly actin, as shown by selective extraction of the A- and I-bands with salt solutions. At high magnification, electron micrographs of muscle in

Fig. 3.3 High power electron micrograph of insect flight muscle in rigor showing crossbridges linking thick and thin filaments. (Courtesy of Drs M. C. and M. K. Reedy.)

rigor reveal connections between the thick and thin filaments in the overlap zone: the so-called crossbridges (Figure 3.3). In muscles fixed under relaxing conditions the crossbridges are not so clearly defined, but they appear to emanate from the thick filament. Examination of the individual filaments, obtained by mechanical disruption of relaxed muscle, confirms the source of the crossbridge. Projections are observed along the length of the isolated thick filament, apart from a central bare zone (Figure 3.4). As discussed later, the crossbridge is an integral part of the myosin molecule and its structure and function hold the key to the mechanism of contraction [see Squire, 1986].

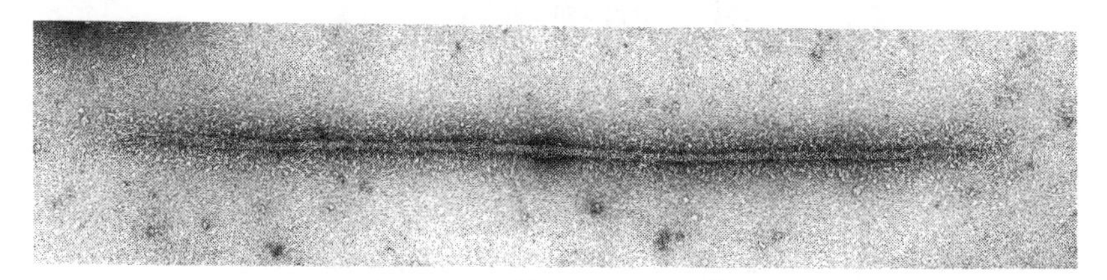

Fig. 3.4 Electron micrograph of a negatively stained thick filament from rabbit psoas muscle showing splayed myosin heads and the central bare zone. (Courtesy of Drs P. Knight and J. Trinick [25].)

3.3 MEMBRANE SYSTEMS

The membrane which surrounds a muscle fibre, the sarcolemma, periodically invaginates the fibre to form T-tubules. Along each side of the T-tubule lie the internal membrane compartments of the sarcoplasmic reticulum, so that in longitudinal section of muscle a triad of vesicles is seen (Figure 3.5). Triads appear at each Z-line in frog sartorius muscle, but in other muscles they may appear at the boundary of the A- and I-bands [6]. As alluded to in section 2.3, the function of the T-tubule is to transmit the action potential into the region of the contractile machinery where it stimulates the sarcoplasmic reticulum to release Ca^{2+}. As a result, the diffusion time for the Ca^{2+} message is minimized. Two proteins of central importance to this excitation–contraction coupling

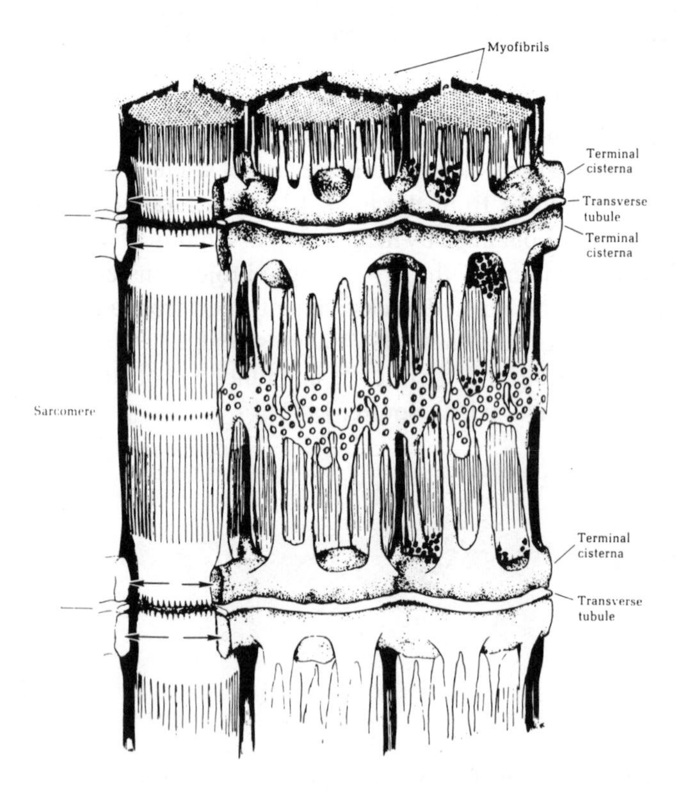

Fig. 3.5 The sarcoplasmic reticulum of frog skeletal muscle. (From Peachey [6].)

have recently been identified. One, blocked by dihydropyridine, is located in the T-tubule membrane, while the other, the so-called ryanodine receptor or foot protein, is located at the junction with the sarcoplasmic reticulum. The mechanism by which the electrical message sensed by the dihydropyridine receptor is passed on to the ryanodine receptor, a Ca^{2+} channel protein, is unclear. It could involve direct coupling through a conformational change or possibly an intermediate chemical message [7]. The ultimate effect, however, is the release of Ca^{2+} from the sarcoplasmic reticulum.

At the cessation of the stimulus, Ca^{2+} is actively pumped back into the sarcoplasmic reticulum. In some muscle types the sarcolemma and mitochondria may also play a role in controlling intracellular Ca^{2+} concentration. When a muscle is blended, fragments of the sarcoplasmic reticulum break off and reseal to form vesicles which can be purified by preparative ultracentrifugation. These vesicles are capable of actively accumulating Ca^{2+} ions in the presence of ATP. The Ca^{2+}-pump ATPase responsible for this action has been well studied [8]. The volume of the muscle cell dedicated to the sarcoplasmic reticulum is roughly proportional to the speed of its contraction. In the toad fish, 25% of the swim bladder is devoted to this structure. The organ is capable of contracting 100 times a second, allowing the fish to emit an audible grunt. However, it remains to be proven that the sarcoplasmic reticulum can accumulate Ca^{2+} with sufficient rapidity to account for relaxation directly. Short term regulation of $[Ca^{2+}]_{free}$ during twitches could be achieved by the action of slow binding, high affinity Ca^{2+} sites as found on the soluble cytosolic protein, parvalbumin.

The direct role of Ca^{2+} in controlling contraction can be demonstrated using permeabilized muscle cells (section 3.4). That Ca^{2+} is the physiological agent responsible for activation of contraction in living cells was confirmed by experiments using the Ca^{2+} indicator, aequorin [see Rüegg, 1986, Chapter 8]. Aequorin is a bioluminescent protein, extracted from a jellyfish, that emits light on binding Ca^{2+}. Stimulation of an intact muscle fibre, injected with aequorin, causes a glow of light whose time course precedes that of tension generation. From the absolute light intensity, the concentration of the $[Ca^{2+}]_{free}$ is estimated to change from about 0.1 to 10 μM, which is comparable to that required to activate a demembranated muscle. The Mg.ATPase activity of myofibrils and

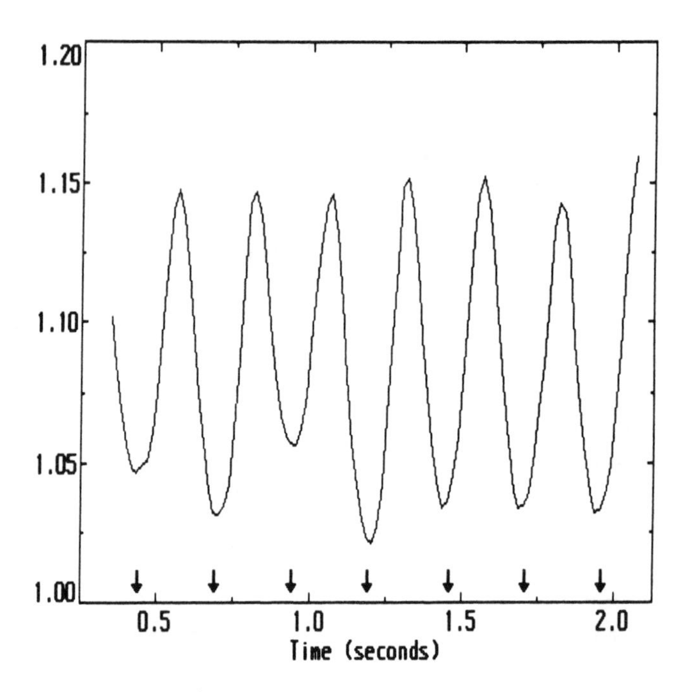

Fig. 3.6 The change in $[Ca^{2+}]$ during the beating of a rat atrial myocardium, electronically stimulated at the times marked ↓, as monitored using the fluorescent indicator, indo-1. (Courtesy of Professor B. Northover.)

crude actomyosin is activated by a similar concentration range of Ca^{2+} and provides the biochemist's probe for regulatory mechanisms (section 4.5).

In vivo $[Ca^{2+}]_{free}$ can also be estimated using fluorescent indicators such as fura-2 and indo-1 [9]. The former shows a wavelength shift in the excitation spectrum. Fura-2, when free of Ca^{2+}, absorbs maximally at 335 nm, whilst the complex with Ca^{2+} absorbs maximally at 362 nm, allowing the $[Ca^{2+}]_{free}$ to be estimated from the ratio of the fluorescence signal when excited at 335 and 362 nm. Indo-1 shows a shift in the emission wavelengths and the $[Ca^{2+}]_{free}$ can be estimated from the emission intensity ratio at 410 and 485 nm. Using a ratio measurement allows the $[Ca^{2+}]_{free}$ to be deduced independently of the sample thickness and indicator concentration. The indicators may be introduced by microinjection, or they can be perfused into the cell as their lipid-soluble esterified forms. The intracellular esterases then generate the ionized indicator which is unable to diffuse back out of the cell. Using video microscopy it is possible to monitor the $[Ca^{2+}]_{free}$ with millisecond time resolution and several micrometres spatial resolution within a living cell. On the other hand, the average $[Ca^{2+}]_{free}$ can be estimated with good sensitivity using a photomultiplier detector. Figure 3.6 illustrates the oscillation in sarcoplasmic $[Ca^{2+}]_{free}$ from beat to beat in isolated heart muscle fibres. Care is needed in such measurements to determine that the signal is not altered by the binding of the indicator to intracellular components.

3.4 FIBRE PREPARATIONS

The membrane system provides a barrier to the bathing medium and prevents experimental manipulation of the chemical environment in the vicinity of the myofibrils. There are several methods for overcoming this problem.

Individual fibres can be mechanically skinned by peeling off the sarcolemma. The key to successful skinning is that the bathing medium must be free of Ca^{2+} ions, otherwise the fibre will be activated and develop local clots of contracted material. Once the preparation is complete, however, the skinned fibre can be induced to develop a steady tension by raising the free Ca^{2+} level. The sarcoplasmic reticulum remains intact but its capacity to accumulate Ca^{2+} is swamped by the bathing medium. In order to control the free Ca^{2+}, EGTA may be added as a buffer in an analogous fashion to pH buffers:

$$pCa^{2+} = pK_{Ca} + \log([EGTA]/[CaEGTA]) \quad (3.1)$$

where pCa is the negative logarithm of the $[Ca^{2+}]_{free}$. The precise value of pK_{Ca} depends on the conditions, but is about 6.

Skinning may also be performed chemically using detergents (e.g. Brij 35) and this method can be applied to bundles of fibres. The first successful permeable muscle preparation was made by immersing a muscle in 50% aqueous glycerol which, by combination of osmotic shock and solubilization, disrupts all the membrane components. The solution also acts as an antifreeze so that the muscle can be stored for many months at -20°C. During this time the soluble proteins and metabolites leach out, leaving a framework of the contractile proteins in a rigor state.

Although addition of ATP to these fibre preparations causes contraction under activating conditions (i.e. $+Ca^{2+}$), the process is limited by diffusion. To avoid this problem in steady-state measurements, the fibre may be preincubated in a solution containing creatine kinase and phosphocreatine, so as to provide an ATP regenerating system within the myofibrils. For transient measurements, ATP can be introduced as an inactive 'caged' precursor and contraction initiated by photolysis of this derivative by a brief flash of light (section 7.4).

Fragments of myofibrils can be prepared by mechanically disrupting a muscle with a blender and purifying by washing at low ionic strength. To prevent precontraction, the muscle is allowed to go into rigor with its ends fixed before blending. Myofibril preparations are convenient for microscopy and biochemical studies but they suffer from the limitation that, in general, their ends cannot be fixed. After a one-off shortening, their structural order is irretrievably lost. In their classic study, Huxley and Hanson [5] took advantage of myofibrils whose opposite ends had fortuitously adhered to the slide and coverslip respectively, so enabling them to adjust the sarcomere length. More recently, apparatus has been designed to record the force generated by a single myofibril [10].

3.5 UNSTRIATED MUSCLE AND NON-MUSCLE CELLS

The striations of skeletal muscle reflect the alignment of the myofilaments within them and provide direct evidence for a sliding filament mechanism of contraction. Is this intricate order essential for the mechanism of sliding? Studies on synthetic actomyosin threads, made by extruding solubilized actin and myosin through a fine needle into low ionic strength solution, suggest not. On addition of ATP a thread shortens, yet the filaments within it are generally disorganized. Indeed when first observed in the 1940s, this phenomenon was taken as evidence that contraction involved gross changes in the lengths of the protein molecules themselves. In the light of the sliding filament theory, it appears that net shortening of the thread occurs, despite a rather random orientation of the filaments, because those which are favourably aligned will contribute to a local shortening and buckling. These observations suggest that the regular arrangement of filaments in striated muscle has evolved to allow rapid and efficient contraction, but this order is not fundamental to the transduction process itself.

Smooth muscle comprises spindle-shaped, mononucleated cells with no obvious cross-striations (Figure 3.7). In the electron microscope, longitudinally running actin filaments are resolved, but myosin filaments are difficult to detect as they tend to depolymerize during processing [11]. *In vitro*, smooth muscle myosin forms filaments, but they readily disassemble under relaxing conditions. It appears that contraction of smooth muscle results from drawing together of actin filaments by myosin filaments. The actin filaments are fixed at opposite ends to structures called dense bodies which are analogous to the Z-line structures of striated muscle.

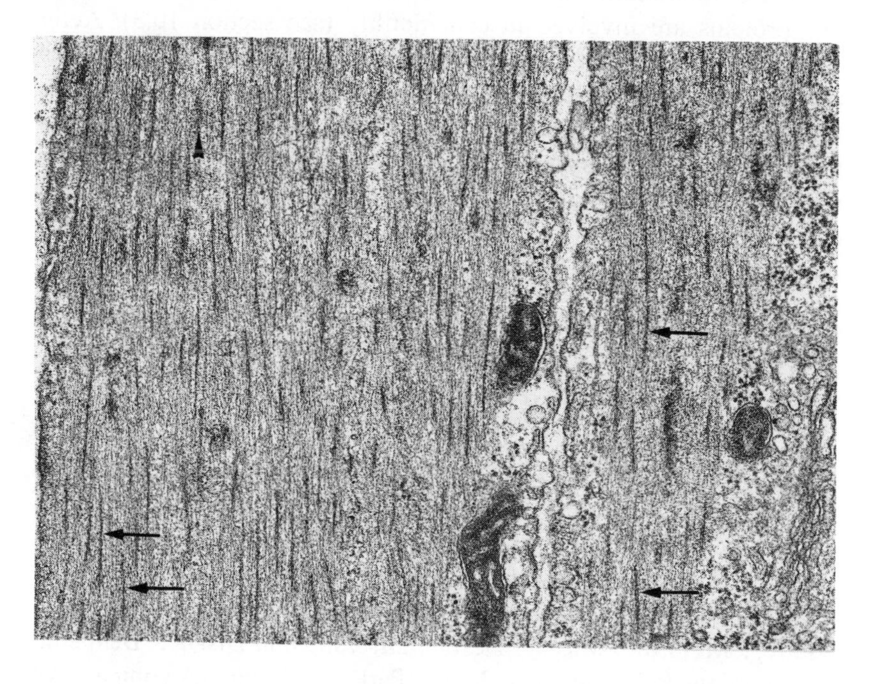

Fig. 3.7 Electron micrograph of the smooth muscle of a rabbit vein. The arrows mark the myosin thick filaments and the arrow head marks a dense body. (Courtesy of Drs A. P. and A. V. Somlyo [13].)

Although this book is concerned with muscle, it is not inappropriate to consider properties of non-muscle cells. Most eukaryotic cells contain actin and myosin, including higher plants, where these proteins play a role in motility, cytoplasmic streaming and shape changes [12]. Under certain conditions actin filaments may be readily observed, such as the stress fibres induced when a fibroblast is plated out onto a substratum. Myosin is more difficult to locate, but it does seem to concentrate at the sites of some contractile activities. During cell division actin and myosin have been identified, by antibody staining, in the contractile ring which constricts to form two daughter cells. Mutant slime mould and yeast cells which lack 'conventional' myosin fail to bud, although the former is still capable of migration, which suggests that alternate myosin or myosin-like proteins are involved in cell motility (see section 10.3). Actin filaments in some plant cells provide a framework on which organelles are transported around the cell, although the accessory motor proteins involved have not been fully characterized. Overall, investigations of these less organized systems indicate that sliding is fundamental to movement and at least one protein must be in a filamentous state. Actin and myosin are much more dynamic in non-muscle cells than in skeletal muscle and filaments are assembled at the sites of action as required [see Bray, 1992].

TOPICS FOR FURTHER READING

Bloom, W. and Fawcett, D.W. (1975) *A Textbook of Histology*, 2nd edn, W.B. Saunders Co., Philadelphia. (Contains a good selection of micrographs of muscle.)

Bray, D. (1992) *Cell Movements*, Garland Publishing Inc., New York and London. (An introduction to cell motility, including muscle.)

Squire, J.M. (1986) *Muscle: Design, Diversity and Disease*, Benjamin Cummings Publishing Co., Menlo Park, California. (A comparative study of muscle structure and function.)

Contractile proteins

4

4.1 PROTEINS OF THE MYOFIBRIL

In order to characterize the protein components responsible for contraction at the molecular level, they must be isolated and purified [see Frederiksen and Cunningham, 1982]. Most purification protocols require a solubilization step which, in the case of myofibrillar proteins, can usually be achieved using high ionic strength solutions. The location of a specific protein within the myofibril may be determined by one of several methods.

1. Microscopy of the remaining myofibril after selective extraction or gene disruption.
2. Comparison by electron microscopy of synthetic polymers of the protein with native structures.
3. Copurification with a major protein whose location is known.
4. Labelling with specific antibodies. Polyclonal antibodies are raised by injection of the purified protein into a different species. The antibodies are isolated from the host's serum and may be labelled with a fluorescent reagent or colloidal gold particles. A myofibril from the original species is then treated with the antibody, which binds to sites (epitopes) specific to the protein in question, and is examined by light or electron microscopy as appropriate. Monoclonal antibodies are prepared from a homogeneous population of cells in culture and recognize a single epitope. However, cross-reactivity between unrelated proteins is occasionally observed because of a chance matching of their amino acid sequence at the epitopic site.

Proteins are characterized with regard to molecular weight and subunit composition using a variety of hydrodynamic and electrophoretic techniques. The analytical ultracentrifuge remains the most important tool for the determination of native molecular weights. Gel permeation chromatography is useful for characterizing molecules on the basis of their size but, under native conditions, the shape of the

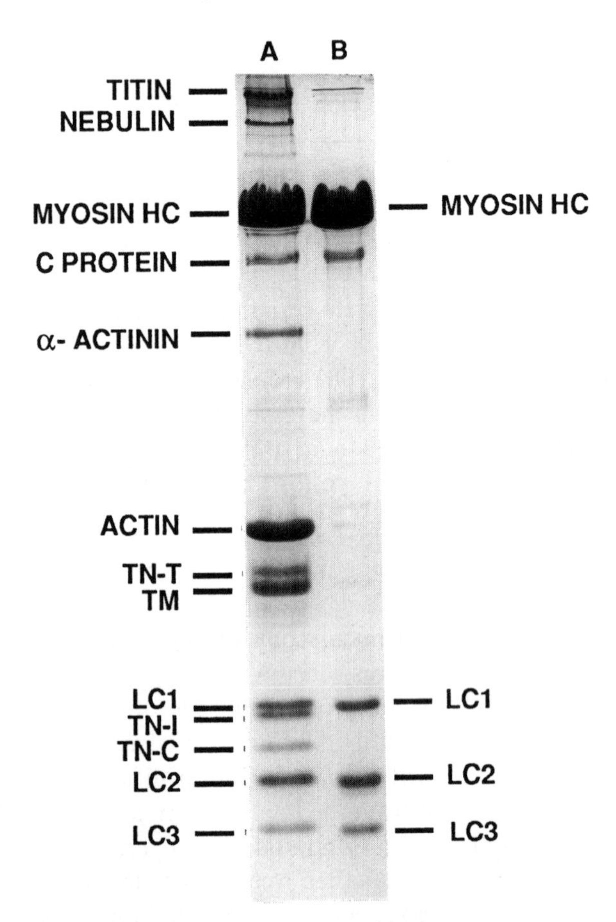

Fig. 4.1 Polyacrylamide gel electrophoresis, in the presence of SDS, of (A) myofibrils and (B) myosin prepared from rabbit skeletal muscle: HC, heavy chain; LC, myosin light chain; TN, troponin; TM, tropomyosin. (Courtesy of Drs D. Swartz and M. Greaser.)

molecule also contributes to the observed elution volume. Electrophoretic techniques can be used to study native proteins but usually they are carried out under denaturing conditions which separate the subunits. Of these methods, separation on a polyacrylamide gel in the presence of sodium dodecyl sulphate (SDS) is the most routine (Figure 4.1). SDS binds to polypeptide chains to give them a similar negative charge per unit length. As the polypeptides permeate the gel matrix towards the anode they separate according to their molecular size. The stoichiometry of the components may be estimated from the relative intensity of staining, but problems can arise from unresolved bands, anomalous stain uptake and adventitious proteolysis.

Myosin (43%) and actin (22%) are the major components of the myofibril [see Ohtsuki *et al.*, 1986]. The regulatory proteins, troponin and tropomyosin, are associated with the actin (thin) filament. The myosin (thick) filament contains a number of minor components, including the M-line protein and C-protein. Invertebrate thick filaments contain a core of paramyosin and in some species this may be the dominant component. Several cytoskeletal proteins have been identified, among them α-actinin, titin and nebulin, which help maintain the structure of the sarcomere (see Appendix, Table A.2).

4.2 ACTIN

Actin is purified by extraction as its monomeric globular form (G-actin) from an acetone-dried powder of minced muscle, using very low ionic strength solutions. On adding salt to physiological concentrations the solution becomes viscous as a result of polymerization to give fibrous F-actin (Figure 4.2). The double helical structure of F-actin is evident from electron microscopy of isolated filaments (Figure 4.11) and X-ray diffraction patterns from actin bundles and intact muscle (section 6.3). G-actin comprises a single polypeptide chain (mol. wt 42 000) which binds Mg^{2+} (or Ca^{2+}) and ATP reversibly. Sequence analysis shows that actin is highly conserved, with only 23 out of 375 residues differing between skeletal muscle and a non-muscle source [14]. On polymerization, the bound ATP is hydrolysed and trapped as non-exchangeable ADP, but this reaction is not essential for the formation of F-actin. *In vitro* at physiological ionic strength, F-actin exists in dynamic equilibrium with

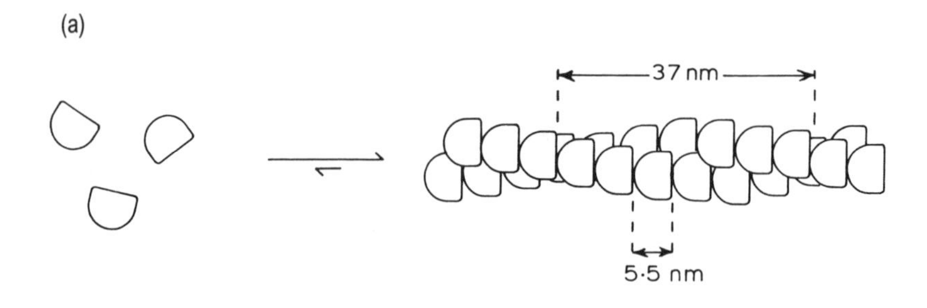

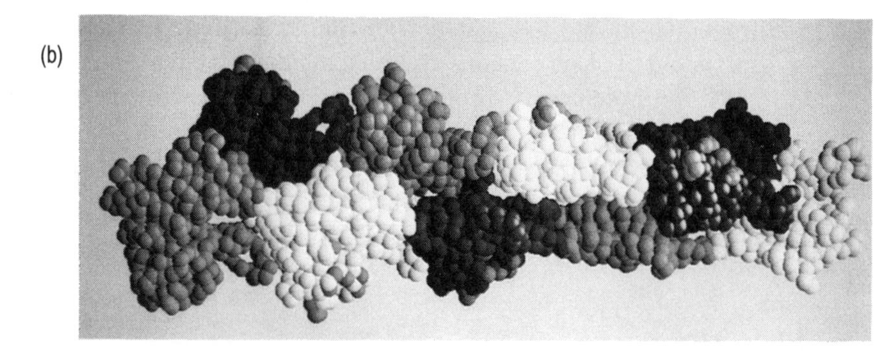

Fig. 4.2 (a) Schematic representation of the polymerization of G-actin to F-actin. (b) Model of F-actin based on the known structure of G-actin. The polarity shown has the 'pointed' end to the right, as defined by S1 decoration. (Courtesy of Professor K. C. Holmes, [17].)

a low critical concentration (about 1 μM) of G-actin. As a result of the continuous addition and loss of G-actin monomers from the ends of the polymer, a low ATPase activity is detectable and may be enhanced by physical disruption of the F-actin by sonication. An F-actin filament has a polarity as a result of the asymmetry of the monomers within it. In the presence of ATP, the relative rates for dissociation and association of monomers at either end of the filament are different, giving rise to the phenomenon of treadmilling.

The polymerization reaction provides one potential mechanism for motility in that the G–F transition is associated with a sol-gel transition of the cytoplasm. An extreme example of this is in the acrosome of some invertebrate spermatozoa whose actin polymerizes explosively on contact with the egg membrane [15]. A second mechanism for generating movement involves interaction with other proteins, such as myosin, which slide along actin filaments in a vectorial manner. Striated muscle presents an extreme example of this where the F-actin of the thin filament is a relatively permanent structure of constant and precisely determined length. The thin filament actin is capped with β-actinin which prevents its depolymerization.

An increasing number of actin-binding proteins have been identified which interact with F- or G-actin, locking it in either a polymerized

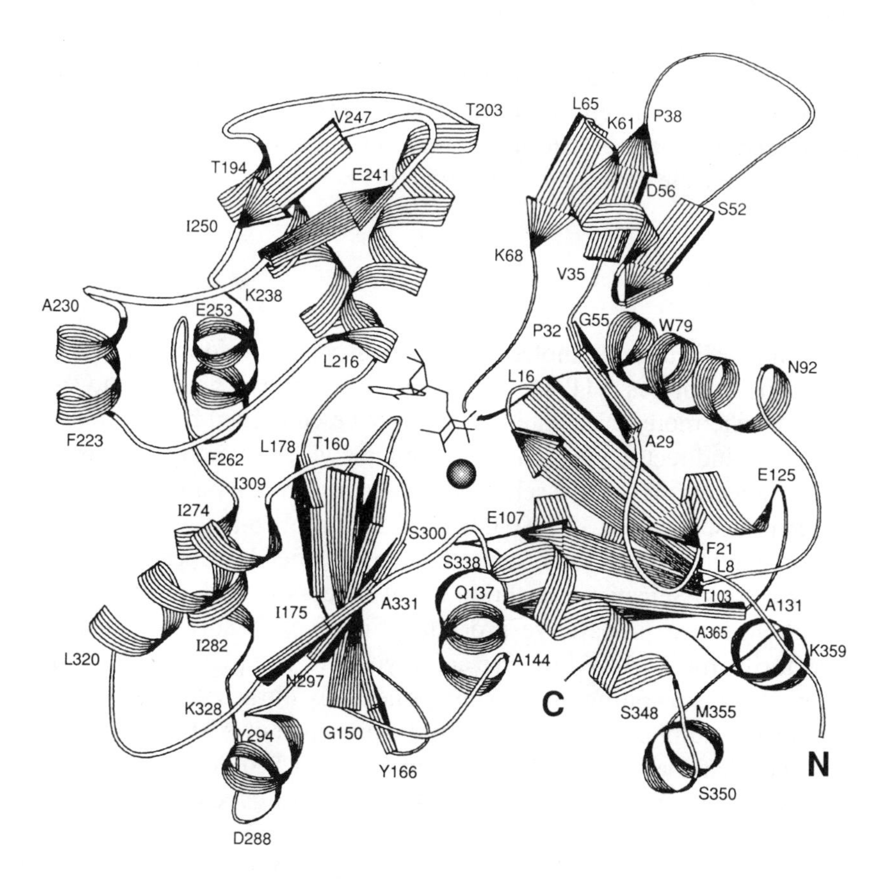

Fig. 4.3 The structure of G-actin determined by X-ray crystallography of the actin-DNase complex. (Courtesy of Professor K. C. Holmes [16].)

or monomeric form. One such protein, DNase 1, forms a tight complex with G-actin which has been crystallized. Analysis of its structure by X-ray diffraction methods has shown that actin has two domains, each containing two subdomains, reminiscent of hexokinase (Figure 4.3). The bound nucleotide is located in a cleft between the two domains [16]. The structure of F-actin has not been obtained at high resolution, although the G-actin molecule can best be fitted into the low-resolution filament structure if the large domain is present on the inside of the double-stranded helix [17]. This conclusion agrees with studies conducted in solution which show that myosin and many other actin-binding proteins interact with the smaller domain. Each actin monomer makes contact with actins above and below it, as well as with those in the opposite strand. The potential for hinge bending between the two domains of actin supports the concept that the actin filament is an elastic structure. The mechanical implications of this may be crucial to muscle contraction (section 10.1).

4.3 MYOSIN

Myosin is an intriguing molecule. It self-associates to form thick filaments, it hydrolyses ATP and it interacts with actin to produce movement. Furthermore, these interactions may be controlled by Ca^{2+} ions, directly or indirectly, via regulatory subunits which are an integral part of the molecule. Myosin is selectively extracted from a mince of rabbit skeletal muscle with 0.3 M KCI, 0.15 M KPi at pH 6.5. It is recovered by precipitation at low ionic strength and is purified by repeat solubilization cycles. Selective extraction is not successful with all muscles. A more general preparative procedure involves dissolving washed myofibrils in 0.5 M KCI to give actomyosin (a complex of actin and myosin) and separating the two components by ultracentrifugation or ammonium sulphate precipitation in the presence of ATP.

Under denaturing conditions myosin (mol. wt 520 000) dissociates into two heavy chains (mol. wt 220 000) and four light chains (mol. wt around 20 000). Early hydrodynamic studies showed that myosin was highly asymmetric and in the electron microscope it is seen to consist of two pear-shaped heads attached to a long tail (Figure 4.4). Each heavy chain forms the bulk of one head and intertwines with its partner to

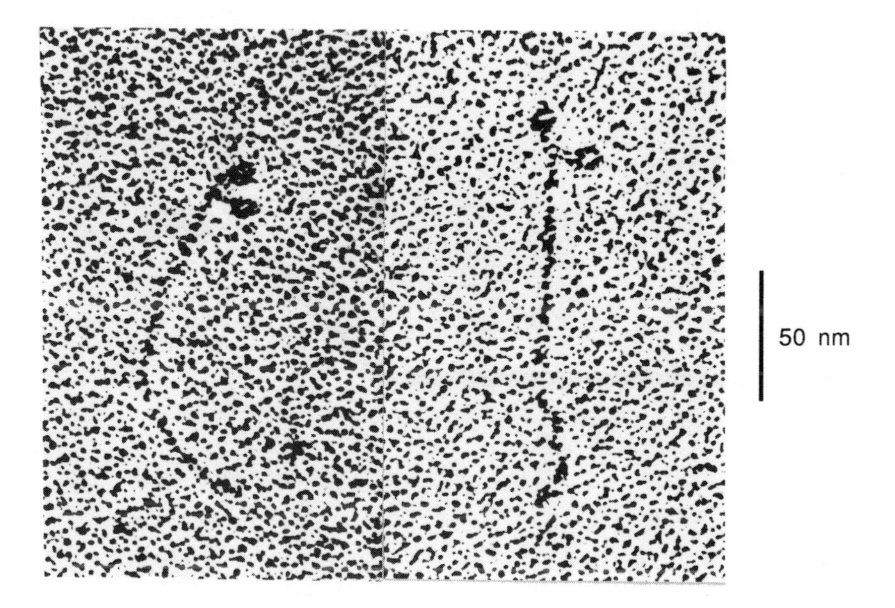

Fig. 4.4 Electron micrographs of rotary-shadowed rabbit skeletal myosin. (Courtesy of Drs A. Elliott and G. Offer [23].)

form the tail (Figure 4.5). Amino acid sequences have been deduced for both the heavy and light chains from a number of species, increasingly through analysis of DNA sequences [see Warrick and Spudich, 1987].

More recently a second kind of myosin has been isolated from non-muscle cells which contains only a single heavy chain. It is termed myosin-I to distinguish it from the conventional double-headed myosin-II described above [18]. The shorter tail of myosin-I does not self-aggregate but instead binds to lipids and also has a secondary actin binding site. It appears to play an important role in cell locomotion and movement of organelles. The head domain has regions of similar sequence to that of myosin-II. The discovery of myosin-I as a viable motor protein suggests that the double-headed nature of myosin-II is not of fundamental importance to mechanochemical coupling. Unless stated otherwise the term myosin refers to myosin-II throughout this book.

Owing to its size, the structure–function relationships of myosin have been unravelled by examining its proteolytic fragments [see Frederiksen and Cunningham, 1982]. Myosin is particularly susceptible to proteoly-

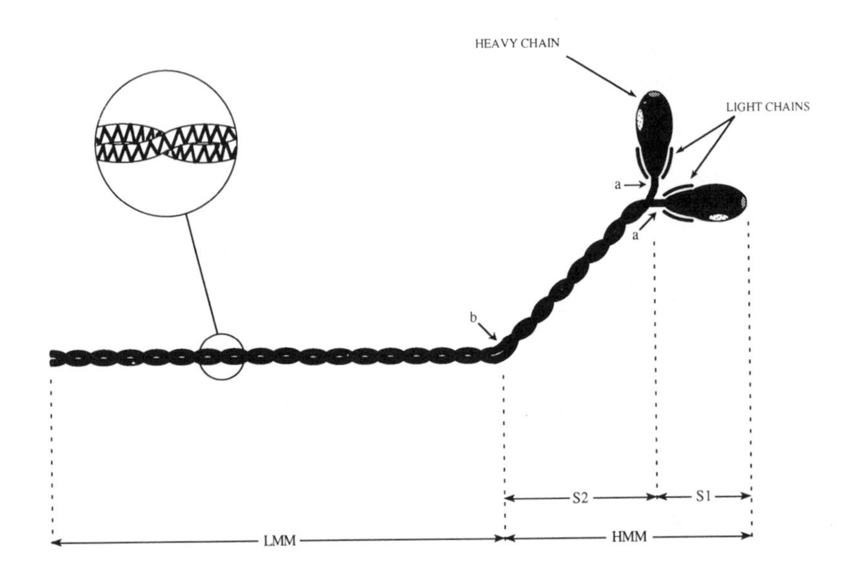

Fig. 4.5 Schematic diagram of the myosin molecule. Proteolytic enzymes attack at the points indicated (a, papain and chymotrypsin and b, trypsin and chymotrypsin) to release the fragments shown. The inset indicates the α-helical coiled-coil structure of the tail.

sis at two locations in each heavy chain (Figure 4.5) and cleavage at one or both sites depends on the enzymes and conditions used. Some key properties of the resultant fragments are given in Table 4.1. These fragments account for the bulk of the parent molecule, but short peptides may be released from susceptible regions and the light chains may be degraded. Limited digestion shows that the ATP, actin and two light chain binding sites are located on each subfragment 1 (S1) moiety and that light meromyosin (LMM) accounts for the self-association of myosin at low ionic strength. X-ray diffraction, optical rotation studies and sequence analysis show that the LMM portion forms a coiled-coil of α-helix [see Squire and Vibert, 1987]. While the tail sequence of myosins are not well conserved with respect to specific amino acids, the arrangement of the different classes of residues is very characteristic. Every alternate third and fourth residue is hydrophobic (the so-called heptad repeat), while between them there are clusters of charged residues. These repeats give rise to parallel stripes in a dot-plot comparison of a myosin sequence with itself or with a myosin from another species (Figure 4.6). In the α-helix (3.5

Table 4.1 Myosin and its proteolytic fragments

Protein	*Symbol*	*Mol.wt*	*Length (nm)*	*Self association*	*Actin and ATP sites*
Myosin	M	520 000	160	+	2
Light meromyosin	LMM	160 000	90	+	0
Heavy meromyosin	HMM	340 000	65	-	2
Subfragment 1	S1	120 000	15	-	1
Subfragment 2	long S2	100 000	60	-	0
	short S2	60 000	40	-	0
Single-headed myosin	SHM	380 000	160	+	1
Rod	rod	260 000	150	+	0

residues per turn) the hydrophobic residues form a continuous stripe along the helix which is responsible for binding to its partner heavy chain. Because the hydrophobic stripe slowly spirals around the α-helix the two chains twist around each other to give a coiled-coil with a pitch of about 14 nm. Charged residues are clustered to give positive and negative patches on the outside of the chains which are responsible for the side by side attraction of staggered neighbouring myosin molecules in the thick filament.

In vitro, at low ionic strength, myosin forms filaments which resemble native thick filaments in having projections along their length and a central bare zone (cf. Figure 3.4). The projections have been identified as the myosin heads since they are lacking in filaments formed from LMM and rods. The presence of a bare zone demonstrates that synthetic myosin filaments are bipolar, but they tend to be of variable length and width compared with native thick filaments. Although the general arrangement of the myosin molecules within the thick filaments is known (Figure 4.7), the precise details have still to be evaluated and vary in different species (section 6.3).

Subfragment 2 (S2) has a similar sequence pattern to LMM and forms a coiled-coil, but it is soluble at physiological ionic strengths. It enables the S1 to project a variable distance from the filament backbone. Note however that a relatively small movement (~10 nm) of S2 away from the thick filament backbone is required for the S1 head (15 nm) to reach the actin surface (see Appendix, Table A.1). Two kinds of S2 have been isolated, namely long S2 and short S2, the latter having lost some labile amino acids at its C-terminus. Correspondingly, long and

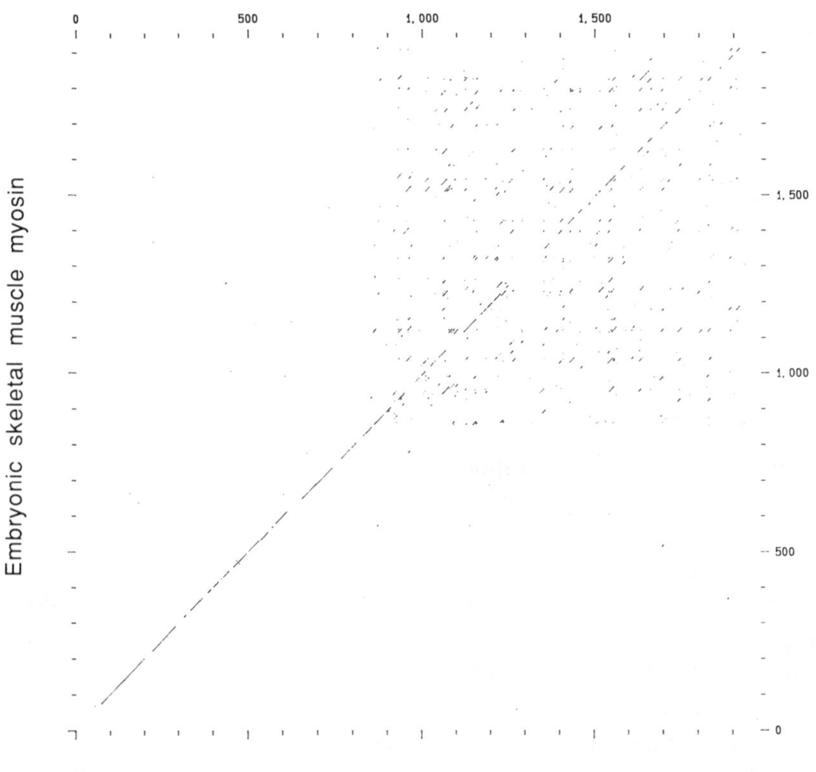

Fig. 4.6 'Dot plot' comparison of the amino acid sequence of embryonic chicken skeletal muscle against chicken smooth muscle. An individual amino acid in one structure is compared, in turn, with all the amino acids in the other and a point placed on the map whenever the amino acids are identical. Identical sequence runs give rise to the diagonal line, as is observed in the N-terminal half (left corner) corresponding to the myosin head. The tail (top right) is less well conserved but the off-diagonal lines reveal the internal similarites of the heptad repeat. (Courtesy of Dr H. Warrick.)

short HMM can also be prepared by the appropriate digestion conditions. Note that the term crossbridge, which originally referred to the connections seen in muscle micrographs (section 3.2), is used somewhat loosely in the literature when applied to the myosin molecule. It may refer to a single S1 head or two heads together with S2.

S1 has been crystallized and solution of its high-resolution structure by X-ray diffraction techniques is imminent [19]. It is evident from electron microscopy of the crystals, that the overall shape of the molecule is that of a bent pear (Figure 4.8). There are indications of a domain-like structure within the head in negatively stained myosin

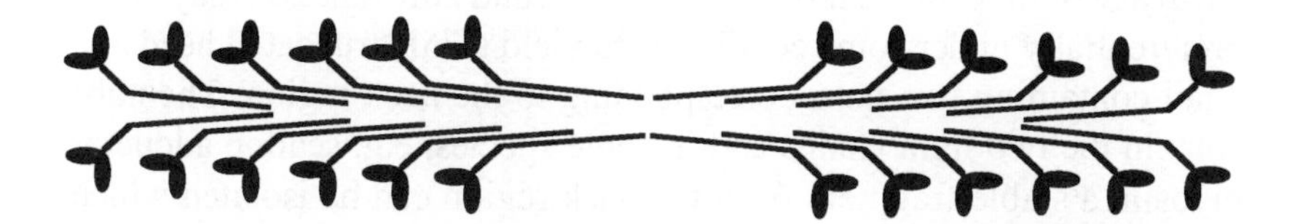

Fig. 4.7 The assembly of myosin monomers to give a bipolar thick filament. The precise packing varies in different muscle types and remains incompletely defined.

molecules [20]. There is also a suggestion of domains from further proteolysis of S1 which, in the case of vertebrate skeletal muscle cleaves to give an N terminal 25K fragment, a 50K fragment and a 20K fragment which is joined to the tail (K in this context represents 1000 = 1 kilodalton).

Other myosins, however, show different cleavage patterns. It has proven difficult to separate these fragments under native conditions for rigorous determination of their functional properties.

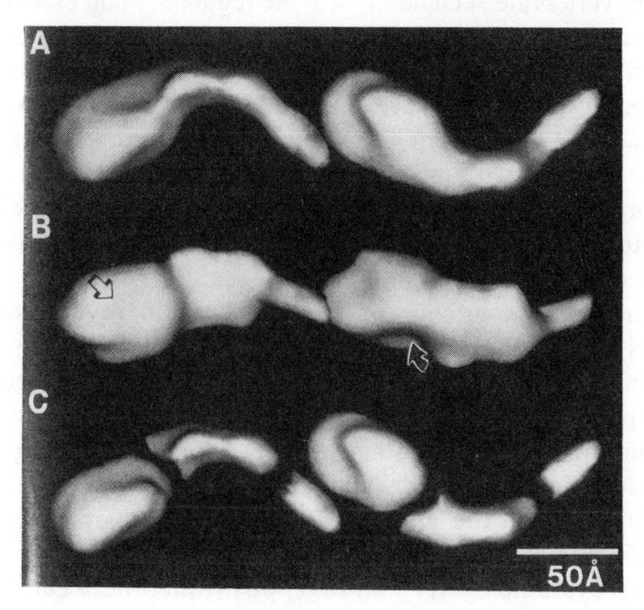

Fig. 4.8 The low-resolution structure of the myosin head determined by electron microscopy of S1 crystals. (A) and (B) show two views of a pair of S1 molecules in the unit cell related by a 60° tilt, while in (C) the molecular boundaries have been limited to reveal the domain structure. The arrows indicate the location of the ATPase sites proposed by Tokunaga *et al.* [28]. X-ray diffraction has revealed a long tract of α-helix, running the length of the molecule. (From Winkelmann *et al.* [19].)

Cleavage of myosin itself between the 50K and 20K junction may occur preferentially under some conditions to yield a 75K truncated head and a tail containing two nubs corresponding to the neck region. The nubs contain the two light chains and in some species, e.g. scallop adductor myosin, a stable fragment from the neck region can be isolated which retains specific Ca^{2+}-binding activity (see section 8.3). Such experiments, along with antibody labelling, confirm that the light chains are located in the neck region of the myosin. Each head contains two light chains which differ in structure and properties, but insufficient is known about them to allow a completely rational nomenclature. Both are members of the Ca^{2+}-binding protein family (section 4.5), although their metal binding functions may be lost. Neither type is required for ATPase activity, but in some species it is clear that they regulate or modulate ATPase activity (Chapter 8).

The nomenclature, regulatory and essential light chains, is still widely used, although it is not a strictly accurate way of classifying them. In vertebrate skeletal myosin the regulatory and essential class of light chains are frequently termed the DTNB and alkali light chains respectively, based on reagents used in their extraction. Three light chain bands (LC1–LC3) are observed in most fast skeletal myosin preparations, when run on SDS-gels, because the alkali class (LC1, LC3) exists in two isoforms (Figure 4.1). Digestion of skeletal myosin, in the absence of divalent metal ions, gives S1 lacking the regulatory light chain (LC2) and the heavy chain is slightly shorter than that of S1 obtained in their presence.

Sequence comparisons of the heavy chain reveal conserved regions within the head of myosins from striated, smooth and non-muscle sources, while the tail is less conserved (Figure 4.6). Using photo-crosslinking methods and by homology with nucleotide binding domains of other proteins, the ATP binding site has been identified with residues around 130 and 170–214 in the 25K domain, although the phosphate moiety may also interact with the 50K domain. Residues are conserved in the region of the reactive cysteines, SH1 and SH2, positioned between 740–750 of the 20K domain. Other conserved regions include the actin binding site(s) located in the 50K and 20K domains (see section 4.4). The sequences in the linker regions between the putative head domains are the least conserved regions and their charge and proteolytic lability suggest that they are exposed to the solvent [see Warrick and Spudich, 1987].

Considerable effort has been directed towards identifying conformational changes in myosin. Addition of ATP causes no detectable change in the amount of secondary structure as assessed by circular dichroism. Changes are observed in tryptophan fluorescence but these may involve local movement of the active site, which occurs in most enzymes on binding their substrates. The reactive cysteines, SH1 and SH2 are in a region of high flexibility as judged by their ability to be crosslinked by reagents more than 1 nm long, but in the presence of ADP they can be linked to each other directly with a disulphide bond [22]. According to fluorescence energy transfer measurements, the reactive cysteines are some 4 nm away from the ATPase site, yet it is clear that these sites communicate with each other.

There is good evidence that the neck and tail regions are flexible at certain points. Proteolytic enzymes of differing amino acid specificity can cleave within the same regions, indicating that the LMM–HMM and S1–S2 junctions are rather open and susceptible to attack at several residues. Furthermore, electron micrographs of myosin preparations show a high proportion of molecules with bends at various angles at these points [23]. For a more quantitative assessment of flexibility, spectroscopic techniques have been applied. These rely on monitoring the change angle of an optical or magnetic probe on the nanosecond to millisecond time scale. Although myosin is intrinsically fluorescent due to its tryptophan residues, better sensitivity is obtained by attaching an extrinsic fluorophore, such as iodoacetylsulphonaphthyl ethylene diamine (IAEDANS), to the reactive SH1 cysteine residue. Rotational motion is assessed by measuring the fluorescence anisotropy decay curve.

Briefly, the S1 is excited by a short flash of polarized light and the polarization of the emitted fluorescent light is followed with time. If the S1, and hence the probe, rotates within the life time of the excited state (about 10 ns), then the emitted light will become depolarized (Figure 4.9). Using this method the rotational correlation time, τ_r, of S1 was determined as about 100 ns, which is consistent with an ellipsoid of 5 nm × 15 nm. Interestingly, the τ_r values for HMM and myosin are only two to three times longer, indicating that the head is free to swing independently about its neck [24]. For the measurement of slower rotations, as is necessary to investigate S1 motion within filaments, the phosphorescence anisotropy decay of eosin-labelled molecules has been used in a similar manner [see Thomas, 1987, Chapter 6].

Electron spin (paramagnetic) resonance (ESR) methods provide similar information but are also capable of resolving different orientational states in steady-state measurements [see Thomas, 1987, Chapter 6]. Here a nitroxide spin label is used which contains a stable unpaired electron (Figure 4.10a). The spin label may be attached to a cysteine residue, such as SH1, or introduced as an ATP derivative. The label absorbs microwaves at specific frequencies dependent on the magnetic field. Three absorption peaks are observed because the electron orbits around the nitrogen nucleus having three magnetic states (+ 1, 0, -1) which add to or subtract from the laboratory field (Figure 4.10b). The first derivative spectrum is usually recorded because the instrument modulates the magnetic field whilst scanning to improve the signal detection. The distance between the peaks, the hyperfine coupling (A), depends on the orientation of the

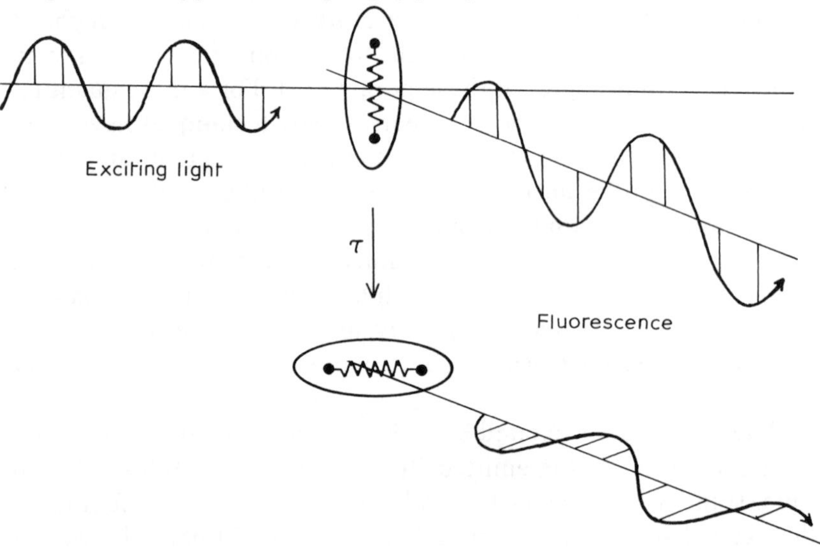

Fig. 4.9 The principle of fluorescence depolarization. Molecules whose absorption dipole is parallel to electric vector of the incoming polarized light are promoted to the excited state, but the polarization of the emitted fluorescence depends on the time of molecular rotation, τ, relative to the life-time of the excited state.

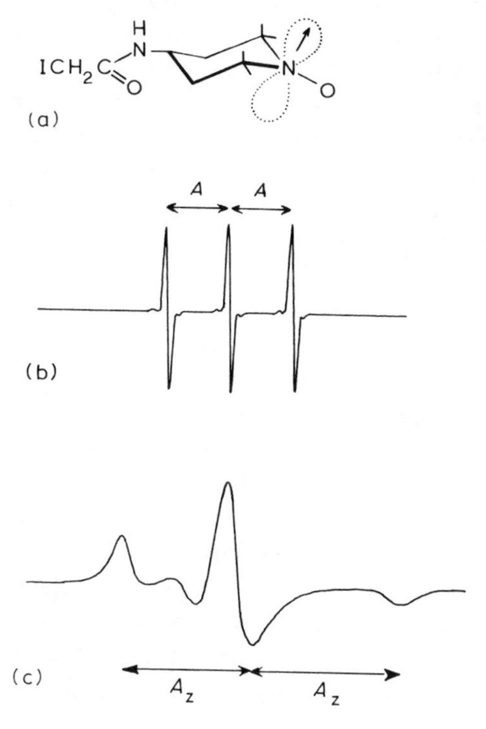

Fig. 4.10 (a) A cysteine-directed nitroxide spin-label containing an unpaired electron in the nitrogen p-orbital. (b) The first derivative ESR spectrum of a nitroxide group in an isotropic environment. (c) A powder spectrum of an immobilized label reflecting the summation over all possible orientations. 'A' defines the hyperfine coupling and is a maximum for those molecules whose orbitals lie in the z-orientation relative to the magnetic field.

label in the field. In solution, the observed ESR spectrum depends on the molecular rotation rate relative to the *A* values expressed in hertz (typically 10^8 s^{-1}). S1 is almost static on this timescale and the SH1 labelled derivative gives a broad spectrum which represents the sum of the signals obtained at all orientations (a so-called powder spectrum; Figure 4.10 c). When ATP is added the spectrum changes indicating slightly increased motion of the label. This is probably a relatively local effect reflecting events within the head, rather than a gross change in its rotational characteristics.

ESR methods have been extended to give information about slower motions. The lifetime of the ESR transition is about 10 μs. At high microwave intensity the population of molecules capable of absorption becomes depleted (saturation). The susceptibility to saturation at a specific frequency depends on whether 'unexcited' spin labels can move from other orientations, within the lifetime of the ESR process, to replenish the supply. Such saturation transfer ESR spectra are sensitive to τ_r values in the range 0.1 ns to 1 μs. This technique has confirmed and extended fluorescence studies on myosin and shows that, even in a myosin filament, the heads are free to rotate through a significant angle on the time scale of 1 μs [see Thomas, 1987, Chapter 6].

These thermally-driven motions are likely to be important in muscle in allowing the myosin head to find an actin site at variable filament spacings [25]. Moreover the LMM–HMM and S1–S2 hinge regions could act as universal joints to transmit motion, developed by movement of the head, to the backbone of the myosin filament.

The rate of ATP hydrolysis by myosin in the presence of physiological Mg^{2+} concentrations is rather slow, but then so is the ATP turnover by resting muscle. At 25°C, the specific activity is about 25 nmol ATP/min/mg fast skeletal myosin, or 0.1 ATP/s/active site. Ca.ATP and K.ATP are hydrolysed at rates of 5 and 20 s^{-1} respectively, but these substrates are of negligible concentration in muscle. Nevertheless, they provide a sensitive *in vitro* assay for myosin during its extraction and purification. Only the Mg.ATPase is activated by actin, therefore this reaction requires more detailed attention.

4.4 ACTIN–MYOSIN INTERACTIONS

In the absence of ATP, actin and myosin form a high-affinity complex corresponding to the rigor state of muscle. Saturation transfer ESR indicates that the myosin heads are rigidly attached and take on the comparatively slow flexing movements of the actin filament ($\tau_r = 1$ ms). Characteristic 'decorated filaments' are seen in the electron microscope when S1 is added to F-actin (Figure 4.11). The arrowhead appearance of the attached S1 confirms that F-actin has a polarity, and indeed has given rise to the terminology 'pointed' and 'barbed' ends.

Decoration with subfragment 1 provides a useful test to identify F-actin in tissues and to determine its polarity within a cell. The polarity of actin is the controlling factor for the direction of filament sliding (see section 9.2). In muscle, the 'pointed' ends of the actin filaments face the

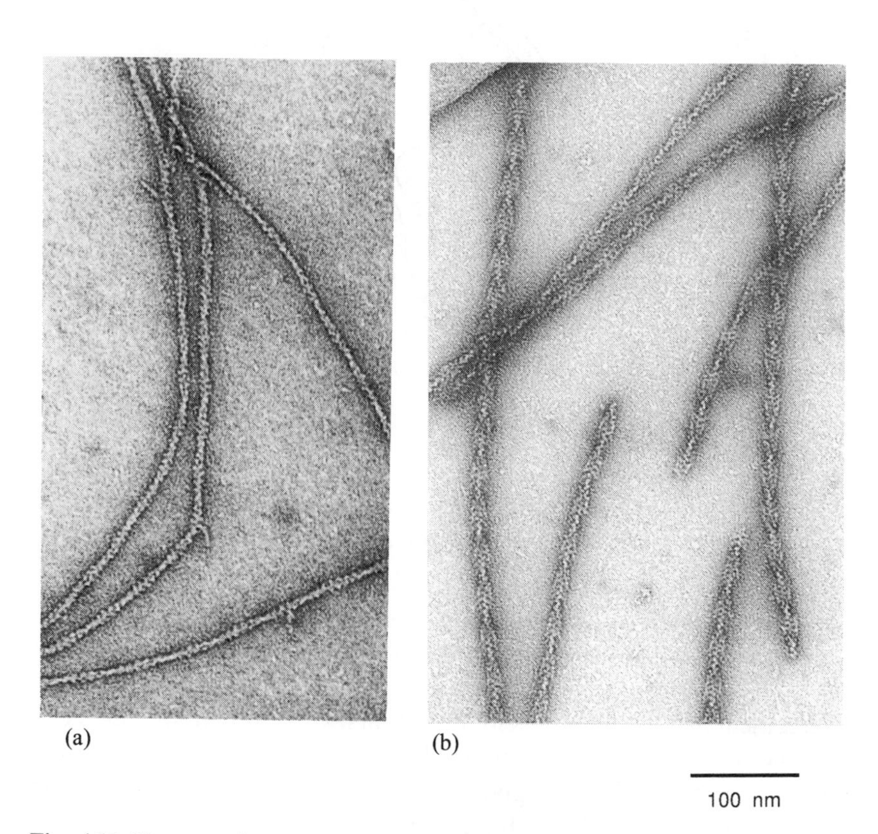

Fig. 4.11 Electron micrographs of (a) F-actin filaments and (b) F-actin filaments decorated with S1 to show the "arrowhead" pattern. (Courtesy of Dr R. Craig.)

centre of the sarcomere. Subdomain-1 of the G-actin monomer, which contains the N- and C-termini, points towards the barbed end of an F-actin filament [17]. NMR and chemical studies have identified the N-terminus of actin (residues 1–40) as constituting a major binding site for myosin, while the corresponding site(s) on the myosin head involve residues 700–720 (in the 20K fragment) and residues towards the C-terminus of the 50K domain (Figure 4.12). Care is required in the interpretation of binding experiments using S1 because the digestion process may expose residues that are normally internal. These residues could bind artefactually to actin as indicated by the ability of S1 to bundle actin filaments over a period of several minutes. Nevertheless, there is general

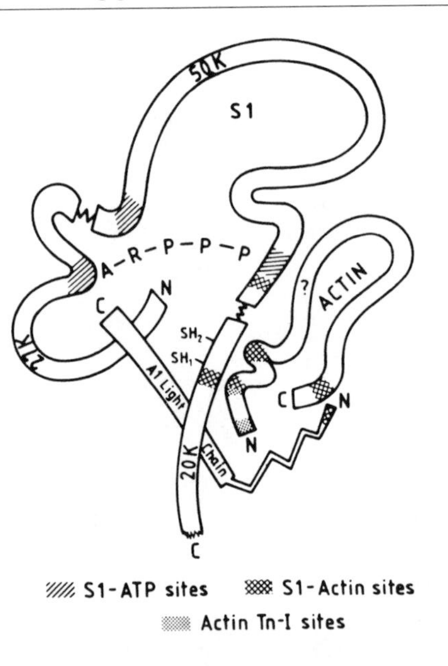

Fig. 4.12 A schematic representation of the primary structures of the myosin head and actin indicating likely points of interaction within their sequence. (From Levine *et al.* [26].)

agreement that the actomyosin interaction involves multiple sites [26].

The appearance of the decorated filament, along with studies of rigor muscle (section 6.2), have given rise to the notion of a 45° attachment angle of the heads. Although this concept is widely used in models of muscle contraction, it should be noted that the arrowhead appearance arises from a two-dimensional view of superimposed heads held on the helical backbone of actin. Nevertheless, this property means that S1 is seen in a variety of known angles and hence it is possible to build up a three-dimensional picture, by computation, using image reconstruction techniques. The method is, however, limited by the distortions introduced in the preparation of the sample for electron microscopy (section 6.2), and the identification of the somewhat arbitrary boundaries of the proteins.

A number of interpretations have been advanced over the last 20 years which assign different regions of the electron density to actin, myosin and regulatory proteins [27]. A recent model, deduced from unstained, rapidly frozen samples, is consistent with the proposed structure for the

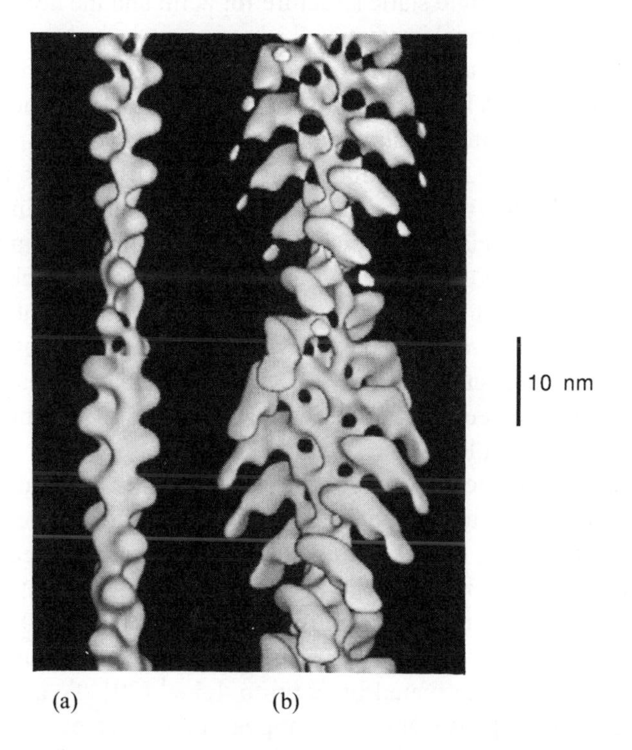

Fig. 4.13 A three-dimensional reconstruction of cryo-electron micrographs of (a) an F-actin filament and (b) an F-actin filament decorated with S1. (Courtesy of Dr R. A. Milligan [27].)

F-actin filament and shows the curved S1 attached tangentially and pointing down at about 42° (Figure 4.13). This study also identified a position of tropomyosin in native thin filaments which has implications for the mechanism of regulation (section 8.2). Using biotinylated probes, it was concluded that the myosin ATPase site (adenosine moiety) is identified as being about 5 nm from the tip of the myosin head and 4 nm from the actin binding site, and about 4 nm from the SH1 reactive cysteine residue, which is located on the opposite side of the head [28]. These studies placed the actin binding site some 5 nm from the SH1 group. On the other hand, biochemical experiments have suggested residues close to SH1 may participate in actin binding (cf. Figure 4.12). This ambiguity should be solved when the crystal structure of S1 is determined. Nevertheless, actomyosin interactions are dynamic and a single static structure for actin and the myosin head may not reveal all potential interactions.

In 0.5 M KCl, actin and myosin form a viscous, turbid solution. On addition of Mg.ATP, the solution clears indicating that the actomyosin complex dissociates. At low ionic strength actomyosin forms a suspension. Again Mg.ATP induces dissociation, but now the ATP is hydrolysed rapidly and, as its concentration falls, the actin and myosin filaments reassociate and contract to form a heavy precipitate (superprecipitation). In early studies this assay provided a quick estimate of ATPase activity; however the complicated steric constraints involved in a two-filament system make the quantitative analysis of actomyosin interactions difficult to achieve. Detailed kinetics studies of solutions have therefore been performed with S1 and heavy meromyosin which remain soluble at low ionic strength.

It is instructive to compare the Mg.ATPase rates deduced *in vivo* (section 2.5) with those measured in solution. The active site concentration in skeletal muscle is 240 μmol kg^{-1} (see Appendix, Table A.1), from which the ATPase activities of resting and isotonically contracting muscle are calculated to be < 0.002 s^{-1} and 6 s^{-1} respectively. The Mg.ATPase of frog skeletal S1 in solution at 0°C is 0.01 s^{-1} and it is activated by actin to 4.5 s^{-1} [29]. To a first approximation, the ATPase activities appear to reflect events in muscle. However, the resting ATPase rate *in vivo* is significantly lower, suggesting additional inhibitory factors may be present. Moreover the comparison of the activated rates is not straightforward (section 7.2) and the numerical agreement might not be as reassuring as it first seems.

4.5 REGULATORY PROTEINS

Although the ATPase activity of crude actomyosin from vertebrate skeletal muscle is dependent on the Ca^{2+} concentration, this property is lost on purification. The Ca^{2+} sensitivity is restored when two proteins, tropomyosin and troponin, are added to the purified actomyosin [30]. Tropomyosin (Tm) possesses the heptad repeat motif similar to that of the myosin tail and thus forms an α-helical coiled-coil dimer. Each tropomyosin molecule extends across seven actin monomers in the thin filament and overlaps with neighbouring tropomyosin by about eight residues to form a continuous, rather flexible structure. Image reconstruction studies indicate that tropomyosin lies in the groove of the actin bound to the inner (large) domain of actin [27]. One troponin is bound to each tropomyosin dimer, the bulk of the troponin lying about 20 nm from the tropomyosin C-terminus (Fig 4.14).

Troponin (Tn) comprises three subunits, designated TnI, TnC and TnT, on the basis of their function (see section 8.2). Structural studies showed that TnC belongs to the Ca^{2+} binding protein family [21]. Parvalbumin was the first member of this family whose crystal structure was solved. Its Ca^{2+}-binding domains comprise a short α-helix, a Ca^{2+}-binding loop, and a second short α-helix. Although specific amino acids are not highly conserved, a characteristic motif is found because the α-helices require hydrophobic residues on their inner faces and the Ca^{2+}-binding loop has oxygen-containing residues, such as aspartic acid, at critical points (Table 4.2). The domain folds around the Ca^{2+} ion in the manner of a clenched right hand, with the extended forefinger and thumb representing the helices and the middle finger forming the loop structure. This domain is termed an E–F hand, based on the structure of the archetypal third domain of parvalbumin [21]. The hydrophobic residues on the inner of face of the α-helices and a conserved isoleucine in the loop results in a stable pairing of two E–F hand domains in many Ca^{2+} binding proteins. The amino acid sequence of TnC indicated that it contained four E–F hands and not surprisingly the subsequent crystal structure revealed two pairs of these domains [31]. However, the crystal structure also showed a long α-helix

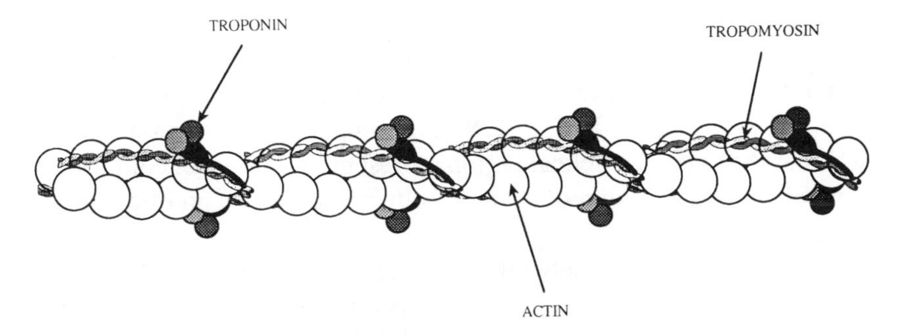

Fig. 4.14 A schematic diagram to show the arrangement of troponin and tropomyosin in the thin filament. Troponin T has an elongated N-terminus which makes extensive contact with tropomyosin. (Courtesy of Dr R. Ankrett.)

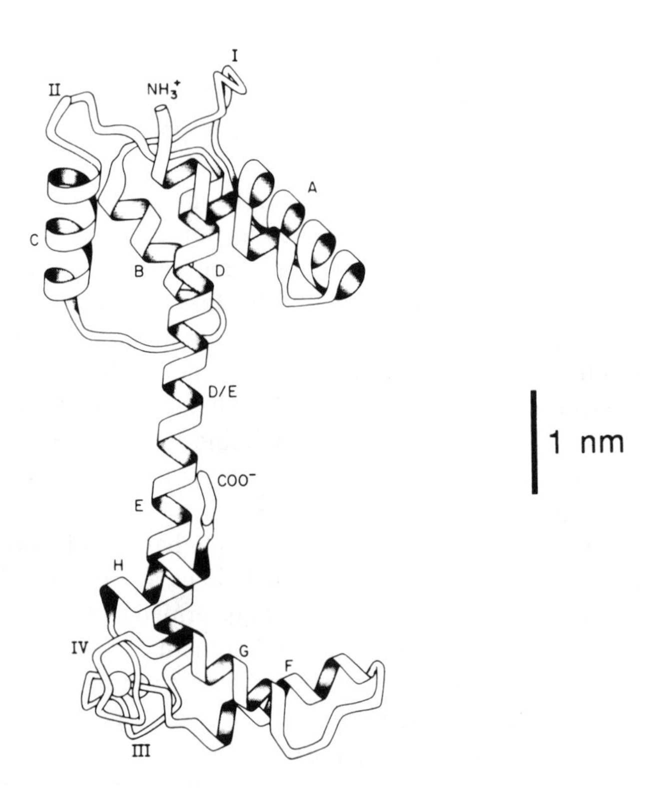

Fig. 4.15 The structure of troponin C solved by X-ray crystallography. (From Herzberg and James [31].)

linking the two halves of the molecule, which was unexpected (Figure 4.15). Its structure is very similar to that of a ubiquitous protein, calmodulin, which is involved in many regulatory processes within cells [see Syrynadka and James, 1985]. It remains controversial as to whether the extended helical region represents a functional conformation when TnC or calmodulin is in solution or part of an oligomeric complex *in vivo*.

TnI is the inhibitory subunit and is rather basic, accounting for its interaction with the acidic TnC. It also binds to actin. TnT is an elongated molecule with a C-terminal head, which interacts with TnI and TnC, and a long tail which extends along the tropomyosin to the Tm overlap region [see Ohtsuki *et al.*, 1986]. The complete troponin

Table 4.2 Amino acid sequence of some EF-hand domains of Ca^{2+}-binding proteins

Parvalbumin (domain 3)	T K T L M A A G	D K D G D G K I	G A D E	F S T L V S E S
Troponin C (domain 3)	L A E C F R I F	D R N A D G Y I	D A E E	L A E I F R A S
Regulatory light chain (domain 1)	F K E A T T V I	D Q N R D G I I	D K E D	L R D T F A A M
Essential light chain (domain 1)	F K E A F L L Y	D R T G D S K I	T L S Q	V G D V L R A L
Critical residues	*h* *h h* *h*	*X* *Y* *Z* *-Y*	*-X* *-Z*	*h* *h h* *h*
	helix	Ca^{2+}-binding loop		helix

h indicates a hydrophobic residue and *X*, *Y* etc. indicate residues which ligand to Ca^{2+}. In the case of *-Y* the backbone carbonyl group bonds to Ca^{2+}.

complex functions to transmit the message from the TnC, on Ca^{2+} binding, to the actomyosin interaction site via a concerted series of allosteric events. Not all striated muscles contain troponin and it is absent from smooth muscles. Tropomyosin, however, seems to be associated with actin in all muscles and is also found in non-muscle cells.

4.6 CYTOSKELETAL PROTEINS

The sarcomere of striated muscle requires other proteins for its assembly and structural stability. These proteins have proven difficult to characterize since they have no enzymic activity and some are so large that they do not enter conventionally-prepared polyacrylamide gels [see Ohtsuki *et al.*, 1986; Trinick, 1991]. α-actinin (dimer mol. wt = 190 000) is located in the Z-line and is responsible for anchoring the actin filaments. Nebulin (mol. wt 800 000) is a giant filamentous protein that runs parallel to the actin filaments, and might function as a length determination factor during assembly. The largest of them all, titin (mol. wt 2 500 000), spans all the way from the Z-line to the bare zone of myosin [32]. It appears to be extensible in the region of the I-band, but is more rigidly bound to the surface of the thick filaments. Titin contributes to the passive elasticity of the sarcomere and helps keep the A-band centralized in the sarcomere (Figure 4.16). Intermedi-

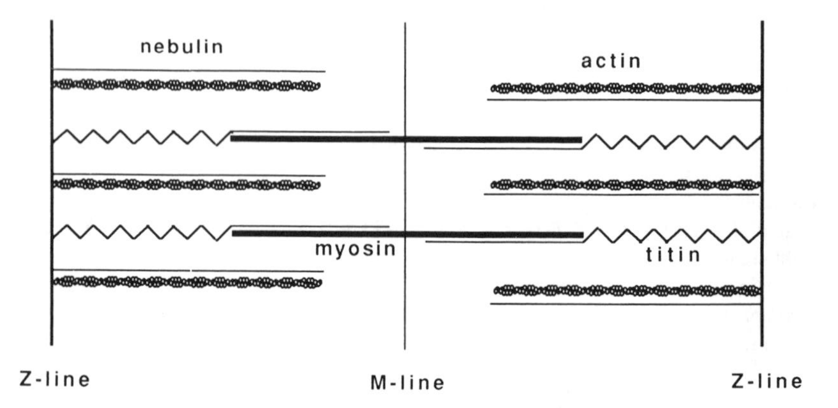

Fig. 4.16 A schematic diagram to show the locations of titin and nebulin within the sarcomere.

ate filaments, so-called because their diameter is between those of thick and thin filaments, link the Z-lines of adjacent myofibrils and help keep them in register across the muscle cell.

TOPICS FOR FURTHER READING

Frederiksen, D.W. and Cunningham, L.W. (eds) (1982) Structural and contractile proteins, Part B. *Methods in Enzymology*, **85**, Academic Press, New York, London. (Covers preparation and characterization of muscle proteins.)

Warrick, H.M. and Spudich, J.A. (1987) Myosin structure and function in cell motility. *Annual Review of Cell Biology*, **3**, 379–421. (Reviews primary structures and functional implications.)

Squire, J.M. and Vibert, P.J. (eds) (1987) *Fibrous Protein Structure.* Academic Press, New York, London. (A structural analysis which includes many of the proteins found in muscle.)

Ohtsuki, I., Maruyama, K. and Ebashi, S. (1986) Regulatory and cytoskeletal proteins of vertebrate skeletal muscle. *Advances in Protein Chemistry*, **38**, 1–67. (Summarizes properties of muscle proteins other than actin and myosin.)

Syrynadka, N.C.J. and James, M.N.G. (1989) Crystal structures of helix–loop–helix calcium-binding proteins. *Annual Review of Biochemistry*, **58**, 951–998. (Reviews troponin C structure and other related proteins.)

Trinick, J. (1991) Elastic filaments and giant proteins in muscle. *Current Opinion in Cell Biology*, **3**, 112–119. (A brief review of cytoskeletal proteins.)

5 Mechanism of ATP hydrolysis

5.1 KINETIC ANALYSIS

In searching for the molecular basis of contraction we need to relate chemical events with mechanical events. These processes must be quantitatively coupled in their extent (thermodynamics) and time course (kinetics). Chemical kinetic analysis of muscle itself presents practical and theoretical difficulties. In a fibre the interaction sites are not freely diffusable and therefore the actomyosin combination process is not a simple bimolecular reaction. Furthermore the rates of some processes will depend on the tension borne by the fibres. It is therefore desirable to start by examining the interactions of isolated proteins because these form homogeneous solutions of defined concentrations and their reactions may be initiated and terminated rapidly. On the other hand, the ATPase kinetic mechanism in solution will necessarily be an incomplete description of events occurring in muscle contraction. In Chapter 7 we consider kinetic approaches for investigating fibres.

Myosin is an enzyme and actin is an activator. The system is therefore amenable to study by standard biochemical kinetic methods [33]. Using catalytic amounts of myosin and a large excess of ATP, steady-state parameters may be deduced from the rate of product formation, v. Actin increases both the V_{max} and the K_m for ATP. There is a loose correlation between the velocity of shortening of different muscle types (V_o) and the V_{max} of the actomyosins extracted from them (e.g. fast-twitch muscles have high actomyosin ATPase activity), although a better correlation with other parameters has been noted (section 7.1). Steady-state methods are useful for routine screening measurements. However, transient kinetic methods are required to gain direct information about the chemical intermediates [34].

The simplest mechanism for an enzyme–substrate (ES) interaction is given in equation 5.1.

$$E + S \underset{k_{-1}}{\overset{k_1}{\rightleftharpoons}} E.S \xrightarrow{k_2} E + P \qquad (5.1)$$

Under steady-state conditions, the concentration of ES remains constant and is negligible compared with the initial [S]. A plot of v against [S] is hyperbolic ($v = V_{max}.[S]/([S] + K_m)$). When the enzyme is saturated with substrate, the product, P is formed at the maximum rate V_{max}, which provides an estimate of k_2 when expressed in terms of μmole product formed/μmole enzymes/s. K_m is experimentally defined as the [S] which gives a velocity of $V_{max}/2$ and for equation 5.1 is defined by $(k_2 + k_{-1})/k_1$. The strength and weakness of such Michaelis–Menten analysis is its generality [33]. Many enzymes fit this scheme, although additional steps must be involved. K_m and V_{max} usually reflect combinations of several rate constants, which can only be unravelled by transient methods. In the latter, high concentrations of E are used so that formation and decay of the ES complex, and other intermediates, can be monitored directly by chemical analysis or spectroscopic means. The time course observed during the transient phase of a reaction is frequently an exponential function and can be fitted to an equation of the type $[X] = [X]_o\, e^{-k.t}$ *or* $[X]_o\,(1\text{-}e^{-k.t})$, where $[X]_o$ refers to the starting (time zero) concentration and k is the observed rate constant. Convenient relationships to note are that the half-time of such a function is given by $t_{½} = 0.69/k$ and the reaction is essentially complete (> 99%) after 7 times $t_{½}$. For equation 5.1, k_1 may be determined by measuring the rate of ES formation under pseudo first order conditions, i.e. by choosing one reactant (e.g. S) to be in large excess so that ES is formed exponentially with an observed rate constant $k_1[S] + k_{-1} + k_2$. When [S] >> Km, k_{-1} and k_2 can be ignored. Another useful approach is to use [E] > [S], to ensure that only a single turnover occurs. Under these conditions the rate of formation of ES provides a measure of k_1[E], while the exponential decay of [ES] provides a measure of k_2 (assuming [E] >> K_m). This approach is valuable because if some of the enzyme is inactive it does not affect the estimate of k_2 (cf. steady-state kinetics where k_2 is calculated from V_{max} assuming fully-active enzyme). Furthermore, in experiments designed to follow the transient phase, the chemical state of S when bound to the enzyme may be analysed. As discussed below for the myosin ATPase, this intermediate is predominantly a long-lived products complex.

For simple schemes, such as equation 5.1, the observed rate constant of a transient phase can provide a measure for the rate constant of one step in the mechanism. However, with branched or multistep schemes the transient phase may become too complicated to solve analytically. Instead the reaction time course can be simulated using computerized numerical integration routines. One method of solving the differential equations associated with each step is to take a small increment of time (δt) and assume a linear relationship to calculate the change in concentration of each species (i.e.$\delta[X] = [X].k.\delta t$). The concentration of X is then redefined as $[X] - \delta[X]$ and the process is repeated for another increment, δt. The simulated trace is then compared with the experimental record and the values for the rate constants (k) are taken, which give the best fit. As with any modelling, an apparently good fit does not prove that the model is correct, although a poor fit suggests that it is inadequate. These same principles of modelling also apply to analysis of the kinetics of force production discussed in Chapters 6 and 7.

In order to apply transient kinetic methods to enzymes, concentration changes of the intermediates must be followed in real time. Several methods have been developed to achieve this in the millisecond time range [see Johnson, 1986]. In the quenched-flow technique (Figure 5.1a), the two reactants (e.g. myosin and ATP) are rapidly expelled from drive syringes, through a mixing jet and driven along a calibrated tube into the quenching agent (e.g. acid). The age of the reaction at the time of quenching is varied by altering the speed of the syringe drive or the length of the tube. The quenched solution is then analysed for the extent of the reaction by chemical means (e.g. HPLC analysis) and a progress curve is constructed. Where a convenient optical signal exists (e.g. a change in absorbance or fluorescence) the progress curve can be measured continuously using the stopped-flow method (Figure 5.1b). Here, after mixing, the reactants pass through a flow-through cuvette into a third stopping syringe. The plunger of the latter hits a backstop, which triggers the recording apparatus (e.g. a storage oscilloscope or computer with a fast

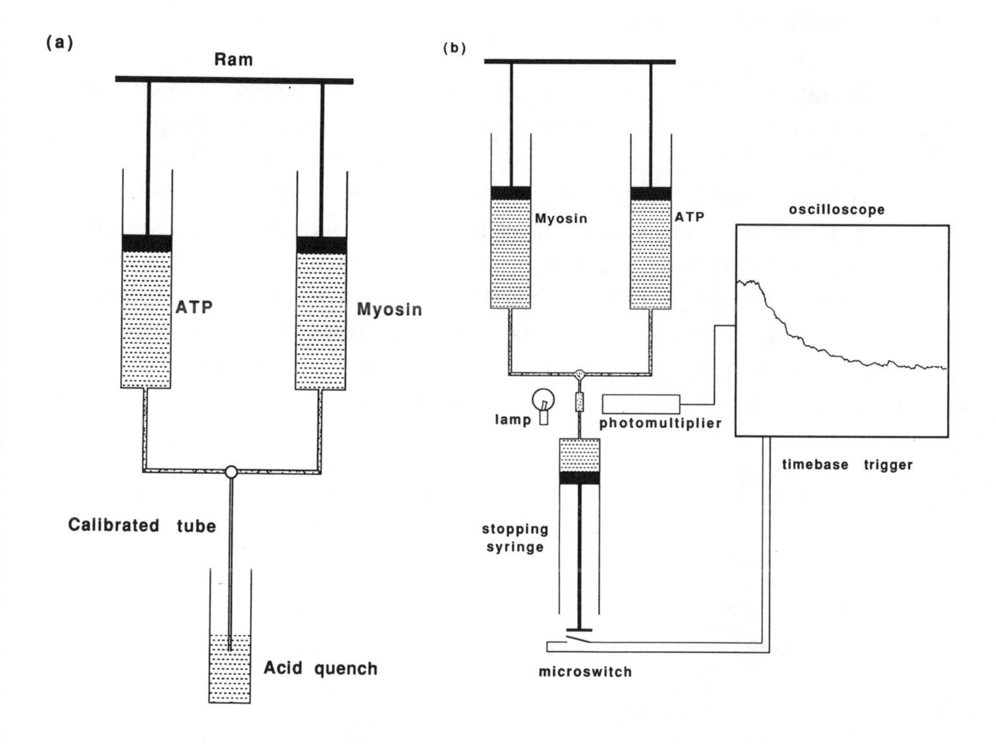

Fig. 5.1 (a) A quenched-flow apparatus. (b) A stopped-flow apparatus for recording rapid changes in absorbance or turbidity.

A/D converter). In this way the optical signal can be recorded as the newly-mixed solutions age in the cuvette. Both of these methods are limited by the time (about 1 ms) taken to mix and transfer the reactants to the quenching or observation chamber. Reactions with rate constants of >1000 s^{-1} will be practically complete within this dead-time of the instrument.

Better time resolution can be achieved using perturbation methods in which a reaction mixture, at equilibrium, is shifted by a rapid change in temperature or pressure and the progress of the reaction is followed as the new equilibrium position is established [see Geeves, 1991]. Perturbation methods are of less general use because of the need to work with reactions having moderate equilibrium constants, so that more than one state is present at a significant concentration. On the other hand, they may reveal reaction steps which are unresolved by rapid mixing methods.

5.2 MYOSIN ATPASE

Steady-state ATPase activity can be measured by a number of different methods. Unfortunately there is no difference in the optical properties of ADP and ATP ($A_{260} = 15.4\ mM^{-1}\ cm^{-1}$), thus precluding a direct assay. However, the H^+ release accompanying hydrolysis (see equation 2.1) can be measured continuously with a pHstat or by including a pH indicator in a lightly buffered reaction mixture. Alternatively, ADP or Pi can be measured using appropriate linked enzyme assays which give absorption or fluorescence signals [see Harris and Bashford, 1987]. For example, the following reactions are coupled to NADH utilization or formation and can be monitored by the absorbance at 340 nm.

$$\text{ADP} + \text{PEP} \xrightarrow[\text{Pyruvate kinase}]{} \text{ATP} + \text{pyruvate} \xrightarrow[\text{Lactate dehydrogenase}]{\text{NADH} \quad \text{NAD}} \text{lactate} \quad (5.2)$$

$$\text{Pi} + \text{G3P} \xrightarrow[\text{Glyceraldehyde 3-phosphate dehydrogenase}]{\text{NAD} \quad \text{NADH}} \text{2,3DPG} + \text{ADP} \xrightarrow[\text{Phosphoglycerate kinase}]{} \text{3PG} + \text{ATP} \quad (5.3)$$

where PEP is phosphoenol pyruvate, G3P is glyceraldehyde 3-phosphate, 2,3 DPG is 2,3-diphosphoglycerate and 3PG is 3-phosphoglycerate. In both reactions ATP is regenerated, thereby maintaining the initial ATP concentration. ATPase activity can also be determined discontinuously by stopping the reaction with acid and analysing the products; ADP by HPLC and Pi by a number of colorimetric procedures. The most sensitive method of detection, however, is to use radio-labelled ATP as a substrate, such as γ^{32}P-ATP.

The analysis of the myosin ATPase mechanism is greatly aided by the production of soluble fragments, S1 and HMM which form low-viscosity solutions at physiological ionic strength. The mechanism described refers specifically to S1 from rabbit fast skeletal muscle at 20°C, although its characteristics seem to be fairly general. Numerical values are given for illustrative purposes and often have been rounded. In any event, precise literature values are meaningless without detailed specified conditions. In the steady state, the methods for ATPase determination described above all give a V_{max} corresponding to 0.1 ATP/S1/s. However

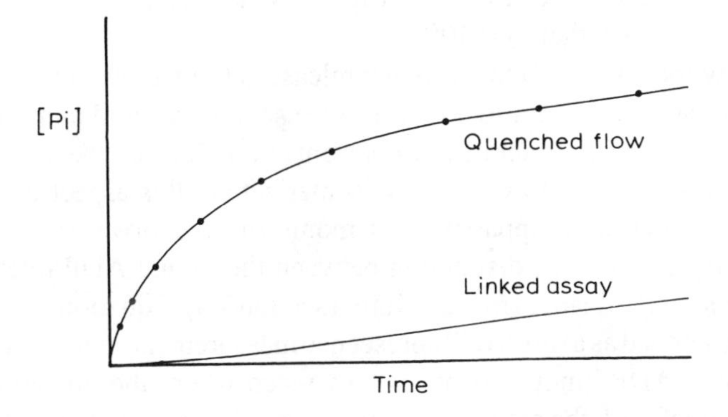

Fig. 5.2 The time course of free and bound phosphate production when a molar excess of ATP is mixed rapidly with S1.

if the time course of product formation is extrapolated back to zero time, the acid quenching procedures reveal a burst phase whose amplitude is almost equal to the [S1]. The time course of this process is resolved using the quenched-flow technique, as shown in Figure 5.2, where the burst of Pi is followed by its steady-state production. The linked-assay system detects only free Pi and does not give a burst. The acid quench must therefore release protein-bound Pi when it denatures the S1. Similar results are obtained if ADP production is monitored. The data therefore indicate that during the first round of interaction, ATP binding is rapidly followed by hydrolysis to yield bound ADP + Pi, and that release of these products is rate limiting in the subsequent steady-state phase. Equation 5.4 is the simplest scheme which accommodates these results.

$$\mathrm{M + ATP} \underset{}{\overset{k_1}{\rightleftharpoons}} \mathrm{M.ATP} \underset{\text{fast}}{\overset{k_2}{\rightleftharpoons}} \mathrm{M.ADP.Pi} \underset{\text{slow}}{\overset{k_3}{\rightleftharpoons}} \mathrm{M + ADP + Pi} \qquad (5.4)$$

where M denotes the myosin head. In the quenched-flow experiment, the burst phase is attributed to the formation of M.ADP.Pi which decomposes in acid. The rate constant for the rapid initial burst of Pi production (k_{obs}, determined from the exponential profile in Figure 5.2) is limited by either the binding step (k_1[ATP], a pseudo first order rate constant) or the hydrolysis step (k_2), whichever is the slower. At low [ATP], k_{obs} is linearly dependent on [ATP] and yields a value for $k_1 = 10^6$ M^{-1} s^{-1}. At high [ATP], k_{obs} is independent of [ATP] and shows that $k_2 = 100$ s^{-1}.

Further work has shown that product release is limited by an isomerization process, subsequent to which Pi release is very rapid and ADP then dissociates with a moderate rate constant (1 s^{-1}). Because these faster product release steps follow the slow isomerization, this aspect of the mechanism is not usually apparent when monitoring the forward kinetics of the reaction. However, distinction between the Pi and ADP release steps becomes significant in muscle fibres (section 7.3). Equation 5.4 has also been expanded as a result of fluorescence measurements which show that ATP and ADP binding is at least two-step where the nucleotide initially collides with the active site to form a weak complex which then isomerizes to a tight complex. For example:

$$M + ATP \rightleftharpoons M^{*}.ATP \rightleftharpoons M^{**}.ATP \qquad (5.5)$$

where * denotes the level of tryptophan fluorescence emission. The isomerization probably involves a conformational change within the protein. Nevertheless, the concentration of M^{*}.ATP is usually so low that it may be ignored for most purposes and equation 5.4 remains sufficient.

So far in the argument we have assumed that the reverse reactions are negligible. Although the overall equilibrium constant for ATP hydrolysis lies far towards products and an enzyme cannot alter this property, some steps may be freely reversible. Thus $K_1.K_2.K_3 = K_{eq} = 10^6$ M (see section 2.5).

To explore the reversibility of step 2 in equation 5.4, consider the following single turnover experiment (i.e. [enzyme] present in excess of [substrate]).10 μM S1 is mixed with 1 μM ^{3}H-ATP so that binding will occur with a pseudo first order rate constant, k_1[S1] = 10 s^{-1}. Hydrolysis will occur immediately (since $k_2 = 100$ s^{-1}) so that the predominant species after 1 s from mixing, before the products are released, will be M.ADP.Pi (see equation 5.4). However, analysis of the radioactive nucleotide from the acid quenched reaction mixture at this point revealed 10% ATP and 90% ADP, suggesting that $K_2 = k_2/k_{-2} = 9$. This remaining labelled ATP, attributed to M.ATP, disappeared with a rate constant of 0.1 s^{-1} showing that it was in equilibrium with the bound ADP during the product release phase. Addition of unlabelled ATP after 1 s had no effect on the decay of the labelled ATP, showing that once bound the only route available was hydrolysis and product release; thus $k_{-1} << k_3$.

The reversibility of the hydrolysis step is confirmed by isotope exchange measurements [35]. If the ATPase reaction is carried out in $H_2{}^{18}O$, then more than one ^{18}O is incorporated into the Pi. Furthermore, addition of high concentrations of ADP and Pi to S1 gives a small, but detectable amount of M.ATP, as detected by quenching and TLC separation [36]. Thus the irreversibility of the overall ATPase reaction is largely reflected in the first step of ATP binding. As a consequence of the negligible value of k_{-1}, the K_m for ATP is given by $k_3/k_1 = 10^{-7}$ M.

The products in the M.ADP.Pi complex are not covalently bound, in contrast to the ion pump ATPases where a phosphorylated intermediate has been isolated. Thus, when this steady-state intermediate is quenched in an acid solution containing $H_2{}^{18}O$, no ^{18}O is incorporated into the liberated ADP or Pi. Further evidence against a phosphorylated intermediate comes from stereochemical considerations. The myosin ATPase reaction results in the inversion of the stereochemical configuration of the oxygen atoms of the terminal phosphate group which suggests that step 2 is a simple in-line attack of water (equation 5.6).

$$\text{R—O—P} + \text{H—O—H} \longrightarrow \text{R—O—P—O(H)H} \longrightarrow \text{R—O} + \text{P—O} \qquad (5.6)$$

ATP H$_2$O transition state ADP Pi

Vanadate, Vi, behaves as a phosphate analogue in many biological systems. Addition of Vi to the myosin in the presence of ADP results in the formation of a stable M.ADP.Vi complex which has similar structural properties to the M.ADP.Pi state. The M.ADP.Vi complex has a lifetime of many hours and provides a useful species for investigation by techniques having a poor time resolution. In the presence of intense u.v. light, the Vi becomes chemically reactive and cleaves the myosin heavy chain at serine residue 180, so establishing one location of the nucleotide binding site in the sequence [37].

5.3 ACTIN ACTIVATION

The myosin ATPase is activated by F-actin filaments, but not by G-actin. Conventional solution kinetics apply in the case of F-actin interaction with S1. However, steric complications are introduced when the heads are tethered as in HMM, and these are further compounded in myosin filaments where the phenomenon of superprecipitation occurs (section 4.4). In practice, the interaction of HMM is similar to S1 in many respects, a finding which is consistent with the high degree of flexibility about the S1–S2 link. Myosin, however, shows a lower degree of activation presumably because the probability of two filaments coming together in the correct orientation is rather low and the head periodicity does not match that of the actin sites. There is considerable controversy as to whether the two heads of myosin function identically and independently. Two heads do not seem to be vital for mechanochemical energy transduction (sections 4.2 and 9.1). The pairing of the myosin heads may simply reflect the requirement for a stable coiled-coil in the rod.

Another problem encountered in kinetic analysis concerns the high protein concentration of muscle (about 100 mg/ml actomyosin) which cannot be achieved in solution. A compensatory, but not completely satisfactory strategy is to measure the ATPase at very low ionic strength (10 mM) where actin has an enhanced affinity for the S1 intermediates. Under these conditions the V_{max} of the actin-activated S1 ATPase is about 20 s^{-1}, the K_m for actin is 40 μM and the K_m for ATP is 10 μM. Actin and myosin are present in locally high concentrations in the myofibril but their study is not straightforward. In the presence of excess ATP, myofibrils tend to supercontract so that only a brief period is available for measurement of their ATPase [38, 39]. Addition of substoichiometric amounts of ATP does allow single turnover analysis to be performed [40]. Another way to mimic the high actin concentration *in vivo* is to chemically crosslink S1 to actin. Such a complex shows a very high ATPase activity, even at physiological ionic strengths where non-crosslinked acto-S1 would be largely dissociated and poorly activated [41]. Note that actin concentrations are expressed in terms of the equivalent molarity of the G-monomers, although significant activation of the ATPase is achieved only with the F-form.

Steady-state kinetic analysis suggested that actin behaved as a simple activator, but the process seemed paradoxical in that addition of ATP to acto-S1 causes the turbidity and viscosity to drop practically to the values expected for the dissociated proteins [see Taylor, 1979]. Actin must therefore achieve activation by a transient interaction with S1, but this is sufficient to bypass the slow product release step limited by k_3. Equation 5.7 represents the potential interactions of actin with the S1 intermediates of equation 5.4.

$$
\begin{array}{ccccccc}
\text{A.M + ATP} & \underset{}{\overset{k_4}{\rightleftharpoons}} & \text{A.M.ATP} & \overset{k_5}{\rightleftharpoons} & \text{A.M.ADP.Pi} & \overset{k_6}{\rightleftharpoons} & \text{A.M + ADP + Pi} \\
k_{-a}\updownarrow k_a & & k_{-b}\updownarrow k_b & & k_{-c}\updownarrow k_c & & k_{-a}\updownarrow k_a \\
\text{M + ATP} & \overset{k_1}{\rightleftharpoons} & \text{M.ATP} & \overset{k_2}{\rightleftharpoons} & \text{M.ADP.Pi} & \overset{k_3}{\rightleftharpoons} & \text{M + ADP + Pi}
\end{array}
\qquad (5.7)
$$

In the absence of ATP, S1 binds tightly to actin with a stoichiometry of one head per actin monomer. The equilibrium dissociation constant, K_a is about 0.1 μM (note the subscript 'a' refers to the nomenclature of equation 5.7; expressed as an association constant its value is 10 μM^{-1}). Thus a mixture of 10 μM S1 and 10 μM actin exists predominantly (90%) as the A.M complex (the rigor state). With this complex in one syringe of a quenched-flow apparatus and ATP in the other, Pi production can be followed as described previously (cf. Figure 5.2). The time course of the transient phase is very similar to that of S1 alone, but the steady-state rate is now increased to about 4 s^{-1} (note $[A] < K_m$ for actin in this example). Increasing the ATP concentration increases the observed rate constant of the transient phase of Pi production, which reaches a maximum of 100 s^{-1} as before.

The significance of this finding is clarified when turbidity changes are followed in a stopped-flow apparatus. At low ATP concentrations, the turbidity of the acto-S1 drops with the same time course as the Pi burst, but the rate constant increases linearly with [ATP] until it becomes too fast to measure (>1000 s^{-1}). In terms of equation 5.7 it is concluded that k_4 has a similar value to k_1 (about 2×10^6 M^{-1} s^{-1}), but once formed, A.M.ATP dissociates to M.ATP ($k_b > 1000$ s^{-1}) before hydrolysis takes place. Thus, at low actin concentrations at least, Pi formation is limited by k_2 as with myosin alone. The tight binding of ATP to S1 is used, in effect, to drive the dissociation of the A.M complex. Actin then reassociates (k_{-c}) to displace the products and achieve an overall activation ($k_6 >> k_3$).

This pathway, as originally proposed by Lymn and Taylor [42] and marked with bold arrows in equation 5.7, provided the basis for a simple, but influential, model of contraction (section 7.1), but it did not consider the reversibility of all the steps in the pathway. The scheme did not clearly define events at high actin concentrations, but assumed that the M.ATP complex had a lower affinity for actin than M.ADP.Pi, so as to drive a cyclic detachment–attachment cycle during each round of ATP hydrolysis. Subsequently it was shown that these complexes had a comparable weak affinity for actin. Furthermore at high actin concentrations, as the steady-state V_{max} was approached, only a fraction of the myosin was bound to actin. In fact, the V_{max} expressed per mole of actin was higher than that expressed per mole of S1, showing that one actin monomer could dissociate from one S1 and activate another within the S1 turnover time. Thus it was not possible to simply assign the value of k_6 to the observed V_{max} of 20 s^{-1}.

Two models were put forward to account for these discrepancies. Taylor suggested that k_5 is somewhat smaller than k_2 and k_6, so that high concentrations of actin inhibit the ATPase [see Taylor, 1979]. Thus V_{max} is reached at sub-saturating actin concentrations where the flux is predominantly via k_2 and k_6. Alternatively, Eisenberg and colleagues [43] introduced additional isomerization steps for the A.M.ADP.Pi and M.ADP.Pi species which were rate limiting (the so-called refractory to non-refractory state transitions). Inclusion of extra states gives a better fit to the kinetic data, but this is generally true for models with increased parameters. However, it does not readily satisfy ^{18}O exchange data which show that at high [actin] the incorporation of extra ^{18}O into the Pi is reduced, in accord with the hydrolysis step becoming rate limiting [44]. These ambiguities are difficult to resolve because at high [F-actin], the solutions become very viscous, preventing accurate measurements. Also, the observed actin affinities are very ionic strength dependent, so that care is required in designing experimental protocols to ensure consistency as the [actin] is increased. As with the myosin ATPase, the Pi and ADP release steps have been resolved and indicate that Pi dissociation occurs first. In the reverse reactions, Pi binding to A.M.ADP is very weak, although isotope exchange studies show it does occur [44].

Overall, it is evident that, unlike the myosin ATPase, a single rate-limiting step cannot be identified for the actomyosin ATPase (i.e. several rate constants have comparable magnitudes). Despite these complications some features of the scheme have received general support. States such as M.ATP and M.ADP.Pi have a low affinity for actin (about 100 μM) while others, such as M.ADP and M alone, have a high affinity (<1 μM). During the ATPase cycle the nature of the bound nucleotide will change, and correspondingly myosin intermediates will alternate between weak and strong binding states. Thus a minimal scheme for the predominant states of the actomyosin ATPase is given by

$$
\begin{array}{ccccccccc}
\text{A.M + ATP} & \underset{}{\overset{k_4}{\rightleftharpoons}} & \text{A.M.ATP} & \overset{k_5}{\rightleftharpoons} & \text{A.M.ADP.Pi} & \overset{k_6}{\rightleftharpoons} & \text{A.M.ADP} & \overset{k_7}{\rightleftharpoons} & \text{A.M} \\
 & & k_{-b}\uparrow\downarrow k_b & & k_{-c}\uparrow\downarrow k_c & & \text{+ Pi} & & \text{+ ADP} \\
 & & \text{M.ATP} & \overset{k_2}{\rightleftharpoons} & \text{M.ADP.Pi} & & & & \\
\text{strong-binding state} & & & \text{weak-binding states} & & & & \text{strong-binding states} &
\end{array}
\tag{5.8}
$$

This concept is important in the interaction cycle between actin and myosin filaments in a muscle. Any model which requires physical attachment of these components during contraction also requires transient detachment in order that the filaments may move relative to each other in accord with the sliding filament theory.

Further analysis of acto-S1 kinetics has resolved a two-step actin binding mechanism of the type:

$$
\text{A + M.N} \rightleftharpoons \text{A–M.N} \rightleftharpoons \text{A.M.N} \tag{5.9}
$$

where N is nucleotide. According to this scheme the first step is independent of the nature of N and has an equilibrium constant of about 100 μM, while the second step is dependent on N. For ATP and ADP.Pi, the second step lies to the left so that the weak binding A-M.N state predominates, while for ADP and no nucleotide (rigor), the second step pulls the overall equilibrium to the right to give the strong binding state A.M.N. The first step shows rapid kinetics of dissociation ($k_{-1} > 1000\ s^{-1}$). The second step is pressure sensitive and fluorescence probes indicate that relative movement occurs within domains of both actin and myosin. This pressure sensitivity has provided a means of comparing actomyosin kinetics in solution with those of muscle fibres [see Geeves, 1991].

A number of ATP analogues have been devised in an attempt to lock myosin in a weak or strong binding state for further characterization by structural methods. The vanadate species, M.ADP.Vi, behaves as a weak binding complex, whereas that containing the non-hydrolysable nucleotide, M.AMPPNP, forms a relatively strong binding species (cf. section 7.5). Substitution of a sulphur atom in the terminal phosphate of ATPγS causes the hydrolysis step to become rate-limiting, so providing a weak-binding state analogous to M.ATP [36]. Strong and weak binding states have also been generated by modification of the reactive sulphydryl groups. Blocking SH1 with N-ethylmaleimide (NEM) causes the myosin head to form a rigor-like link which is resistant to dissociation by ATP. Cross-linking SH1 and SH2 with N,N′ p-phenylenedimaleimide (pPDM) generates a species which shows permanent weak-binding properties [45].

TOPICS FOR FURTHER READING

Harris, D.A. and Bashford, C.L. (eds) (1987) Measurement of ligand binding to proteins, in *Spectrophotometry and Spectrofluorimetry; A Practical Approach*, IRL Press, Oxford, Washington. (Includes chapters on equilibrium binding and rapid kinetic methods applicable to contractile proteins.)

Johnson, K.A. (1986) Rapid kinetic analysis of mechanochemical adenosine triphosphatases. *Methods in Enzymology* (ed. R.B. Vallee), **134**, 677–705. (Describes stopped and quenched flow techniques.)

Taylor E.W. (1979) Mechanism of actomyosin ATPase and the problem of muscle contraction. *CRC Critical Reviews in Biochemistry*, **6**, 103–164. (Discusses the relative merits of alternative kinetic models.)

Geeves, M.A. (1991) The dynamics of actin and myosin association and the crossbridge model of muscle contraction. *Biochemical Journal*, **274**, 1–14. (Reviews the properties of weak and strong binding states.)

Crossbridge structure and function

6

6.1 WHAT MAKES FILAMENTS SLIDE?

There have been many hypotheses for the mechanism of contraction [see Ingels, 1979; Huxley, 1980]. Some sink without trace; none of them have a smooth ride. Prior to the evidence reviewed in section 3.2, one of the favoured mechanisms involved shortening of the myosin filaments, an idea spurred on by the then recent discovery of the α-helix and its prevalence in myosin. When an α-helix melts to a random coil, the ends tend to come closer together for entropic reasons. There is no doubt shortening can be achieved by this mechanism. Meat shrinks when cooked, myosin molecules shorten when heated. The question is deciding whether this is a contributory force *in vivo*. The constancy of the A-band width and the lack of any large changes in the high-angle X-ray diffraction pattern, diagnostic of the coiled-coil structure, shows that the bulk of the myosin tail does not melt during contraction. Filament sliding seems to be well established as the basis of shortening. These studies, however, do not rule out a more localized melting as being a significant event.

Many other hypotheses of contraction, particularly those which postulate long-range charge interactions, have been inspired by the impressive order of a striated muscle. Indeed thick and thin filaments have been likened to capacitor plates which, when oppositely charged, have a natural tendency to shorten to maximize charge interaction. The actual driving force only occurs at the ends of the filaments which bring new positive and negative charges together, so this idea predicts that the tension would be independent of overlap and drops to zero at complete overlap (cf. Figure 6.1). Alternatively it could be imagined that the filaments develop like charges, which result in repulsion. Coupled with the constant volume behaviour of intact muscle this force would cause a shortening. However, skinned fibres do not adhere to the constant volume relationship yet they generate practically the same tension as intact fibres, so indicating that lateral expansion is not a causal phenomenon. None of these ideas have found particular favour with biochemists, as more and more biological interactions are unravelled which illustrate that macromolecules communicate by a specific molecular recognition involving complementary surfaces rather than long-distance telepathy [26]. On the other hand, this kind of intuitive reasoning convinced their predecessors that muscle must contract by a melting helix because myosin is so rich in α-helix.

The current, most widely accepted theory involves so-called independent force generators [see Huxley, 1980], which have become synonymous with the crossbridges observed between actin and myosin filaments. It has remained buoyant for more than 30 years because it is the one for which physiologists, structuralists and biochemists can pool their ideas and produce testable models. The theory explains the sliding of filaments by the action of an array of individual elements within the filaments which could produce a force in the direction of shortening. The idea was born before crossbridges were identified as physical entities in electron micrographs, and indeed draws on physiological evidence from the 1920s. In 1957, A.F. Huxley [46] published an influential model which accounted for the steady-state force, velocity and energy expenditure of a muscle (Figures 2.4 and 2.7). A ‘side-piece’ elastically connected to one filament was proposed to attach to the other filament in a strained state with a moderate rate constant (denoted as f), but to detach rapidly, under the influence of ATP, only when in an unstrained or compressed state. The strained state was assumed to arise from thermal motion, but ATP hydrolysis was essential to provide a net vectorial reaction. The coupling between muscle shortening and ATP hydrolysis (Figure 2.7) arises from the variable rate constant for detachment (denoted as g). When the filaments are fixed, isometric tension is generated by the attached strained side-pieces which accumulate because of their unfavourable dissociation rate, Consequently, the cycling rate and the

ATPase rate is less than in an isotonic contraction where the side-pieces can move into an unstrained position. In a modification to the theory, the decrease in ATP utilization at high velocities of contraction (Figure 2.7b) was explained by a two-step attachment process. The first step was postulated to be freely reversible so that many crossbridges do not complete their cycle before detaching. In effect, contraction is thought to be driven by a few side-pieces whose thermal energy happens to lie at the upper end of a Boltzmann distribution. This does not defy thermodynamics because such high velocities are only achieved at low loads, where the actual work output approaches zero (Figure 2.4). The maximum shortening velocity, V_o, occurs when the force from the positively strained side-pieces just exceeds the negative force from compressed side-pieces. The Huxley 1957 model is a kinetic one but, soon after its proposal, crossbridges were seen in the electron microscope emanating from the thick filament. These entities were correlated with the head of the myosin molecule and thus were identified as the site of ATP hydrolysis. The crossbridge was therefore taken to be the side-piece which undergoes cyclic attachment [47]. The identity of the elastic element(s) remains in question.

If crossbridges do act as independent force generators, then the net isometric tension should depend on the number attached and hence it should be proportional to the degree of overlap between the thick and thin filaments (cf. the maximum interaction hypothesis above). While the expected correlation between tension and length of an intact muscle is observed to a first approximation, a quantitative comparison is complicated by passive elasticity and non-uniformity of sarcomere lengths [48]. These problems were reduced by examining the central portion of a single muscle fibre whose length was controlled by a feedback circuit (section 6.5). When corrected for the passive elasticity present in the unstimulated fibre, the initial isometric tension developed is in good agreement with that expected from the degree of overlap (Figure 6.1). The tension does continue to creep, however, and examination by light microscopy shows that the sarcomere lengths become uneven. An isolated muscle can contract beyond the point of maximum overlap (supercontraction), although the tension is reduced, presumably for steric reasons. The ends of the actin filaments first overlap to form a visible C_m contraction band and then the thick filaments buckle at the Z-line to give a C_z contraction band.

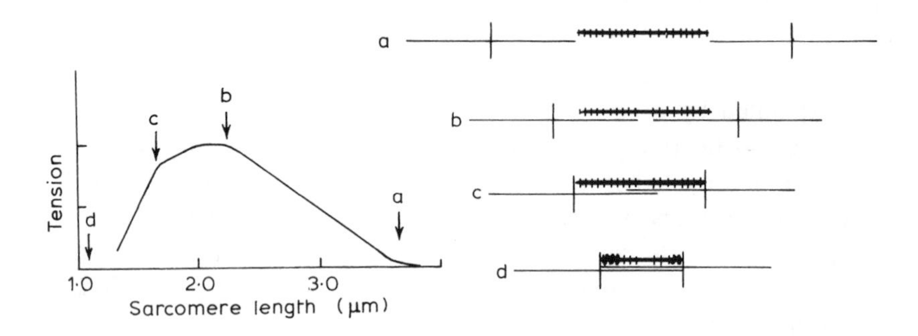

Fig. 6.1 The dependence of the active tension generation by a frog muscle fibre on the sarcomere length. The passive elasticity of a relaxed muscle rises steeply above a sarcomere length of 2.5 μm and this has been subtracted from the total tension of the stimulated muscle. The positions indicated correspond to sarcomere lengths of (a) 3.65 μm, (b) 2.2 μm, (c) 1.65 μm, (d) 1.05 μm. (From Gordon *et al.* [48].)

The negative length–tension relationship at partial filament overlap suggests that there would be a disastrous instability problem. If one sarcomere shortened slightly more than another, it would develop more tension and continue to shorten at the expense of its neighbour. Two factors help to prevent this happening. In the short term, the sarcomere undergoing stretch will develop more transient tension to nullify the change [61], as indicated by the discontinuity of the force-velocity curve at negative velocities (section 2.4). Secondly, there is a passive elasticity which comes into play at long sarcomere lengths which resists over-stretching. This resistance comes partly from the membrane and surrounding connective tissue and partly from cytoskeletal proteins within the myofibril, such as titin (section 4.6). Nevertheless, isolated fibres are particularly prone to creep when maintaining isometric tension and in demembranated fibres only a limited number of contraction–relaxation cycles can be achieved before local clots of supercontracted material appear. It has been found that allowing the muscle to undergo a brief unloaded shortening, followed by restretching, between isometric contractions, prolongs the life of a fibre *in vitro* [49].

A second finding in support of independent force generators is that the maximum velocity of contraction is independent of the degree of overlap and hence of the number of potential interaction sites available in each sarcomere [see Huxley, 1980]. It follows from this relationship that a muscle with a large number of short sarcomeres can contract faster, but develop less tension than one with fewer, longer sarcomeres. While this correlation is found in some arthropod muscles, other factors, such as the myosin isoform, may be involved. Finally, recent *in vitro* motility assays (Chapter 9) have demonstrated movement using isolated myosin molecules, adding strong evidence for their functional independence. However although the independent force generator concept provides a viable mechanism for filament sliding, it reopens the question as to the nature of force production on a smaller scale.

6.2 ELECTRON MICROSCOPE STUDIES

The problem of muscle contraction has become one of defining the structure and movement of the crossbridge relative to the actin filament. Electron microscopy has played a key role in the investigation of this problem. Early studies, which exploited the high degree of regularity of insect flight muscle, revealed that the crossbridges appeared to project outward at about 90° in the relaxed state, with a periodicity of 14.5 nm [50]. This spacing reflects the packing of the myosin molecules in the thick filament (Figure 3.4). In rigor, the crossbridges attach to the thin filament at an angle of about 45° and appear bunched at intervals of 38.5 nm, which probably reflects the position of suitable actin sites (Figure 4.2). These observations led to the idea that the crossbridges might attach at some angle around 90°, then move to 45° so causing the filaments to slide relative to each other by a few nanometres [51].

Can we synthesize a 'movie' of the crossbridge cycle from a collection of electron micrograph 'stills'? There are a number of formidable problems. During contraction the crossbridges move rapidly and asynchronously and only the most stable states (e.g. rigor) can be trapped readily. Furthermore, the processing of the sample for electron microscopy leads to distortions which are not easily separated from any natural disorder. Image processing methods can filter out random aspects of the micrograph, but this may also remove genuine features of the sample [see Squire, 1981]. Nevertheless, techniques such as optical filtration have been widely used to enhance periodic features of muscle specimens. The principle of the method is shown in Figure 6.2.

The electron micrograph (o) is illuminated with parallel monochromatic light and an image (i) is formed with the lens (l). Each point on the

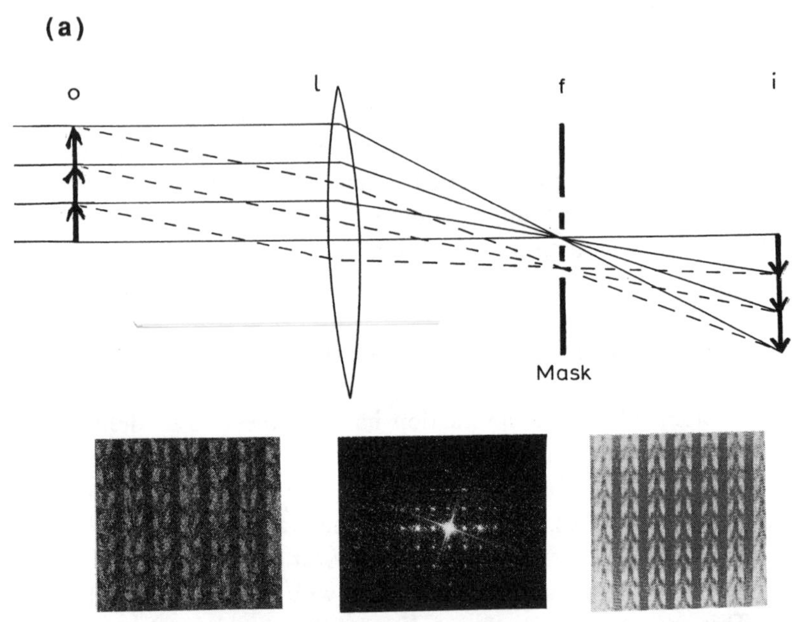

Fig. 6.2 (a) The principle of optical filtration (see text). (b) Optical diffraction pattern and the corresponding filtered image of an electron micrograph of insect flight muscle. (Courtesy of Drs M. C. and M. K. Reedy.)

micrograph acts as a secondary light source from which a new wavefront emerges. The wavefronts complement each other in the forward direction and in a conventional ray diagram we depict this as a parallel beam which, after refraction by the lens, passes through its focal point. Periodic features of the micrograph cause wavefronts to come into step at certain other angles, represented by the diffracted ray (— —) in Figure 6.2a. Diffracted rays converge at the focal plane (f), but off-axis, to give a set of bright spots (the diffraction pattern). Light scattered by aperiodic features results in a diffuse disc at the focal plane. To filter the image, a mask is made by piercing holes in a foil sheet corresponding to the diffraction pattern. With the mask in place, the background scatter from the aperiodic features is eliminated and regular features of the micrograph are enhanced. Note that an unambiguous image cannot be retrieved from the diffraction pattern alone because phase information is lost. The process of optical filtration can also be carried out by computer methods. Optical filtration of electron micrographs clearly reveals the chevron pattern arising from the crossbridges in insect flight muscle (Figure 6.2b).

This pattern compares favourably with thin filaments that have been decorated with vertebrate S1 *in vitro* (Figure 4.11b).

Conventional processing of muscle sections for electron microscopy involves several steps; fixation by crosslinking agents, dehydration, embedding in resin, sectioning using a microtome and staining with electron-dense heavy metals such as uranyl acetate and lead citrate. All of these processes may cause deterioration of the sample and the fixation process cannot be used to trap transient events. For better time resolution the sample can be rapidly frozen by liquid nitrogen which can fix the surface layer of the sample within a few milliseconds. The sample can be dehydrated at low temperatures by the process of freeze-substitution. This technique, when applied to rabbit psoas muscle fibres undergoing isometric contraction, indicated that most of the heads were attached, as in rigor, but they differed from that state in showing a stronger 14.3 nm repeat. It appears as though the head may bend in the neck region to accommodate the myosin periodicity [52]. More recently it has been possible to look at unstained sections directly whilst still in the frozen state. The images are of low contrast, since they rely on a phase effect from defocusing (cf. early light microscopists, section 3.1) and they are examined at very low electron doses to reduce specimen damage. In principle these images should give the best preservation, although there is still some distortion [53]. Ultimately this approach may successfully trap crossbridges during contraction without the problems incurred during conventional fixation and staining.

6.3 X-RAY DIFFRACTION STUDIES

The problem with electron microscopy arises from the need to examine non-functional samples that have undergone a number of preparative treatments. In principle an X-ray microscope would be better because it could reveal atomic detail continuously and non-destructively. Unfortunately, because of the difficulty of refocusing X-rays after diffraction by the object, the technique has not been developed to its full potential. However, as we have seen (Figure 6.2a), a diffraction pattern contains the information required to define an image. For periodic structures which give rise to discrete spots, an image can be reconstructed mathematically provided the position, intensity and phase of each spot is known

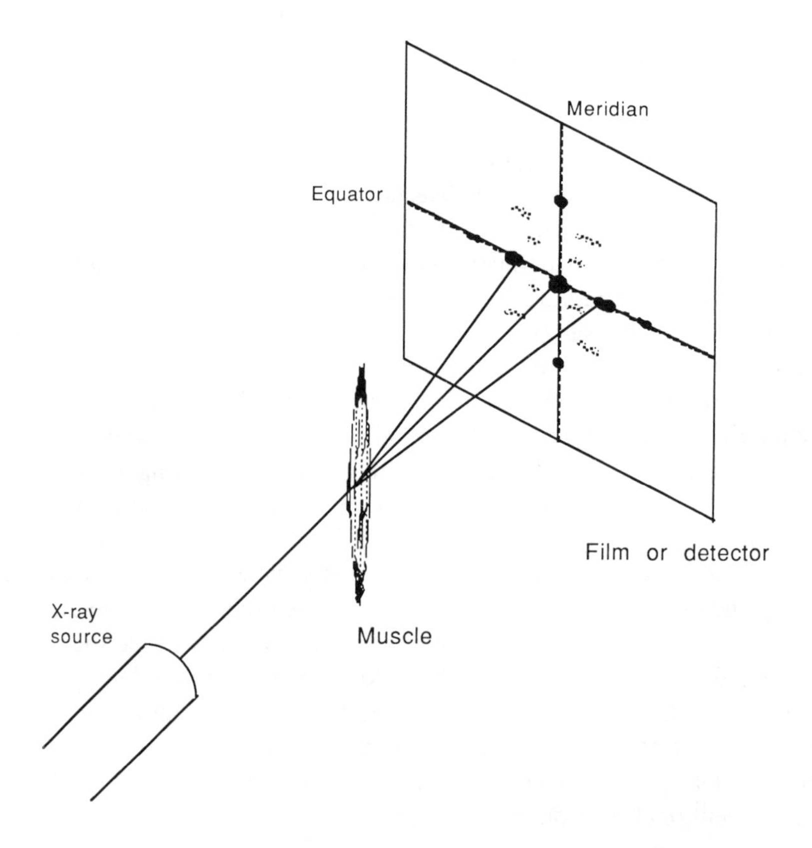

Fig. 6.3 Low angle X-ray diffraction by a muscle. In reality the diffraction angle is only about 2°, but this has been exaggerated in the diagram.

[see Squire, 1981]. Phase determination is a problem. The breakthrough which allowed the high resolution structures of protein crystals to be solved was the introduction of a heavy atom as a reference point [16, 31]. For a simple pattern, without phase information, a structure can be guessed and its computed diffraction pattern compared with the experimental.

Although not as ordered as a crystal, muscle gives a characteristic diffraction pattern which enables periodic structures to be investigated. At high angles of diffraction (corresponding to < 2 nm spacing) the predominant pattern arises from the α-helical rod of myosin and this shows little change on contraction (see section 6.1). At lower angles of

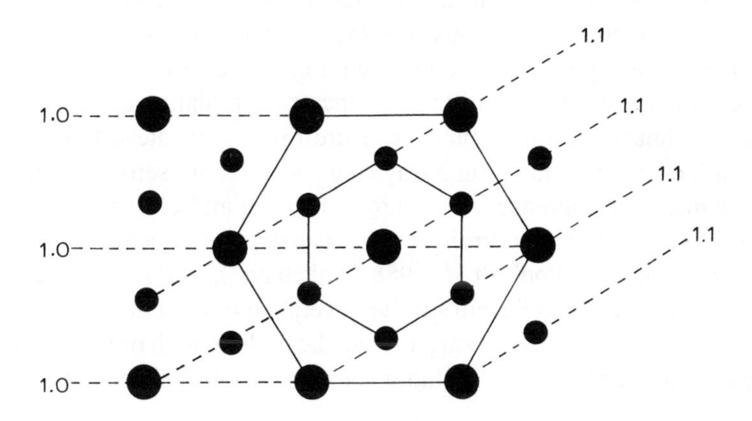

Fig. 6.4 Cross-section of vertebrate skeletal muscle showing the double hexagonal arrangement of the thick and thin filaments. The major equatorial X-ray diffraction pattern arises from spacing between the 1.0 and 1.1 planes. Different muscle types show a considerable variation in the ratio and arrangement of thick and thin filaments.

diffraction, information is revealed about the periodicities of the filaments and their projections. Some of these patterns do change on contraction. Considerable effort has therefore been directed towards identifying the structures responsible for each spot, by correlation of their spacings with periodic features observed in electron micrographs of muscle or isolated filaments and from paracrystals (ordered filament bundles) of purified muscle proteins. Figure 6.3 shows the arrangement for recording a low angle diffraction pattern from a muscle.

The two main spots on the equator arise from the 1.0 and 1.1 diffraction planes of the hexagonally arranged thick and thin filaments (Figure 6.4). Their position therefore monitors the lateral expansion of muscle on shortening. More revealing are their intensity changes. On going from the relaxed to the rigor state the 1.1 spot increases and the 1.0 spot decreases. Since the 1.0 plane reflects electron density of the thick filaments alone, while the 1.1 plane reflects density of both thick and thin filaments, this observation shows that the electron density (presumably crossbridges) moves from the thick filament to the thin filament.

Contracting muscle is more difficult to examine because it can develop isometric tension for only a few seconds before fatiguing. Although diffraction is effectively an instantaneous process, the proportion of X-ray photons that are scattered is small. In practice the time resolution is controlled by the need to accumulate a detectable signal. The first patterns of contracting muscle were obtained by synchronizing the exposure of a photographic plate with excitation of the muscle and a discernible pattern was built up by repeated stimulation over a period of several hours. The minimum exposure times have steadily decreased with increasingly powerful X-ray sources and more sensitive detection systems. Using intense synchrotron radiation and electronic position-sensitive detectors, the stronger peaks can now be detected within a few milliseconds [see Poole *et al.*, 1988; Wakabayashi and Amemiya, 1991]. A frog sartorius muscle will survive several hundred contractions, which enables the X-ray diffraction pattern to be built up with millisecond time resolution by repeated stimulation and synchronized detection [54].

$$\text{Time resolution} = \frac{\text{minimum exposure time}}{\text{number of repeated contractions}} \tag{6.1}$$

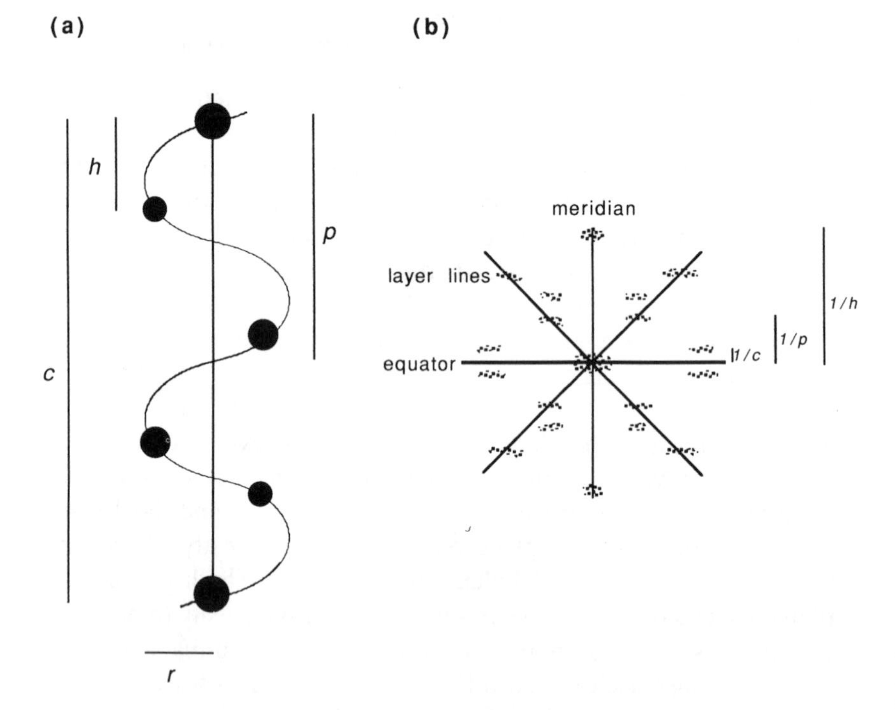

Fig. 6.5 (a) A non-integral helix in which $c = 2p = 5h$, i.e. there are 2.5 subunits per turn. (b) The corresponding diffraction pattern.

The changes in the intensities of the 1.0 and 1.1 reflections indicate that crossbridges move out towards the thin filament a little ahead of tension generation and their final ratio, under isometric conditions, approaches that of the rigor state. The fraction of crossbridges attached, taking these intensities at face value, is between 40 and 80%. The calculation is inexact, however, because the disposition of electron density depends also on the shape and angles of the attached crossbridges. It is also assumed that the detached crossbridges take up the same average position as in the relaxed condition (cf. section 6.6). Although the 1.0 and 1.1 intensity changes provide good supportive evidence that crossbridge movement is a prerequisite for contraction, these equatorial spots are only sensitive to lateral movements. In the crossbridge model outlined in section 6.1, it was assumed that crossbridges move longitudinally. For supporting evidence we need to examine the meridional spots and layer line diffraction pattern. To assign these reflections we must consider the form of the diffraction pattern given by electron density centred at points on a helix [see Squire, 1981]. The example depicted in Figure 6.5 is a non-integral helix where the repeat in mass distribution does not coincide with a single turn of the helix.

The distance of the first layer line from the equator provides a measure of the true repeat, *c*, of the structure, while the distance of the first layer line to appear on a 'cross' passing through the origin gives the pitch, *p*. For an integral helix $c = p$. The distance of the layer line intensity from the meridian contains information about the radial distribution of mass, *r*. The first spot to appear on the meridian provides the subunit repeat distance, *h*. The interpretation of the diffraction patterns from intact muscles is more complicated because several helical structures are involved and these are positioned on a regular lattice [see Squire, 1981].

Relaxed skeletal muscle shows a meridional spot at 14.3 nm and layer lines at 42.9 nm which arise from the helical arrangement of crossbridges about the thick filament (Figure 6.6). The pattern cannot be interpreted unambiguously but it indicates that the crossbridges are arranged on *n* helical strands with a pitch of $n \times 42.9$ nm and an axial repeat of 14.3 nm. The latter is close to the repeat in the projections observed in electron micrographs of thick filaments (cf. Figure 3.4). The weight of the evidence favours $n = 3$ for vertebrate skeletal muscle

(Figure 6.7). In particular vertebrate thick filaments are prone to fray into three subfilaments indicating a three-fold symmetry in packing of the backbone [55]. The interpretation of the X-ray patterns in terms of a helical arrangement does not imply that the myosin tails are twisted together in a triple helix but merely describes the favoured position of the crossbridge in time and space. Invertebrate muscle thick filaments have a larger diameter and n is generally > 3 (e.g. $n = 4$ for insect flight muscle, $n = 7$ for scallop striated adductor muscle). However, the crossbridges display a similar axial spacing (14.5 nm). Note also in striated muscles from invertebrates the arrangement of actin around the thick filament also differs from that shown in Figure 6.4.

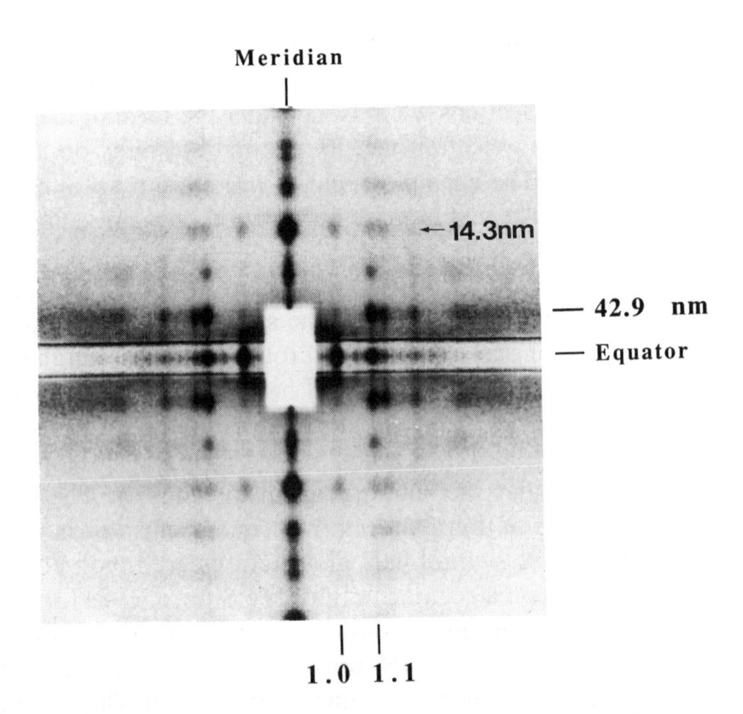

Fig. 6.6 The X-ray diffraction pattern from relaxed fish muscle recorded with an area detector. The diffraction pattern has been weighted to show the weaker outer reflections, while the exposure of the equator has been reduced to display the 1.0 and 1.1 reflections. Note that a spot is labelled according to the real space of the diffracting object (i.e. the muscle), but it appears at a reciprocal position in the pattern. (Courtesy of J. Harford and J. M. Squire [see Squire, 1990, Chapter 10].)

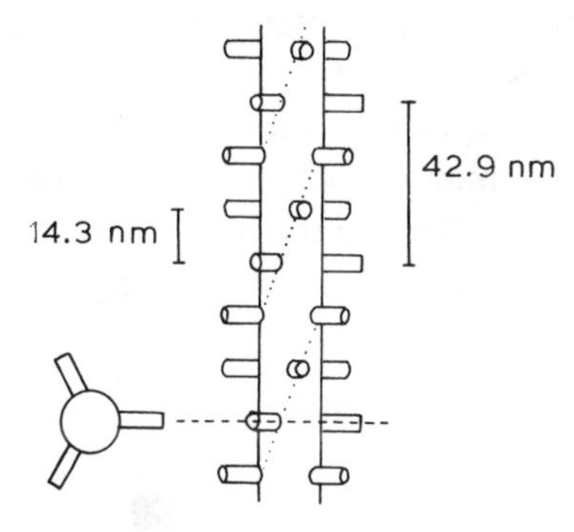

Fig. 6.7 A model of a vertebrate skeletal muscle thick filament in which the crossbridges lie on a three-stranded helix. (From Offer [111].)

While the 14.3 nm spot is the strongest meridional reflection in the vertebrate muscle pattern, the presence of other 'forbidden' meridional spots (e.g. 21.5 nm) shows that the crossbridge spacing is not perfectly regular (Figure 6.6). Nevertheless, the X-ray data indicate that in relaxed muscle the position of the crossbridges is determined essentially by their attachment to the thick filament backbone. This contrasts markedly with the situation in rigor where the crossbridge takes on the spacing of the thin filament.

With the help of actin paracrystals and electron micrographs [17], the helical nature of the thin filament has been established in detail (Figure 4.13). The diffraction pattern, with a meridional spot at 2.7 nm and layer lines at 5.1 and 5.9 nm, arises from a double strand of G-actin monomers with a diameter of 5.5 nm, in which the chains are staggered by half a subunit (i.e. 2.7 nm) with respect to each other (Figure 4.2). The two chains are twisted together with a long pitch of about 74 nm and therefore cross over at 37 nm to give a corresponding layer line. The long pitch of the actin helix is somewhat variable in length. In the double-stranded actin model we can trace out tightly coiled right- and left-handed helices (known as genetic or primitive), which take in alternate staggered subunits of each chain. These genetic helices give the 5.1 and 5.9 nm layer lines.

When a skeletal muscle goes from the relaxed to the rigor state, the 5.1, 5.9 and 37 nm layer lines are all intensified, while the 42.9 nm layer line, attributed to the myosin, disappears. These results indicate that the crossbridges bind and take on the helical characteristics of actin at the expense of losing their thick filament helical symmetry. However, the 14.3 nm meridional is reduced but not lost, showing that crossbridges can find actin sites by a radial (lateral) movement accompanied by a slewing about the azimuth, rather than extensive axial (longitudinal) movement (Figure 6.8). Despite the mismatch in the actin and myosin periodicities and the variable lateral spacing at different sarcomere

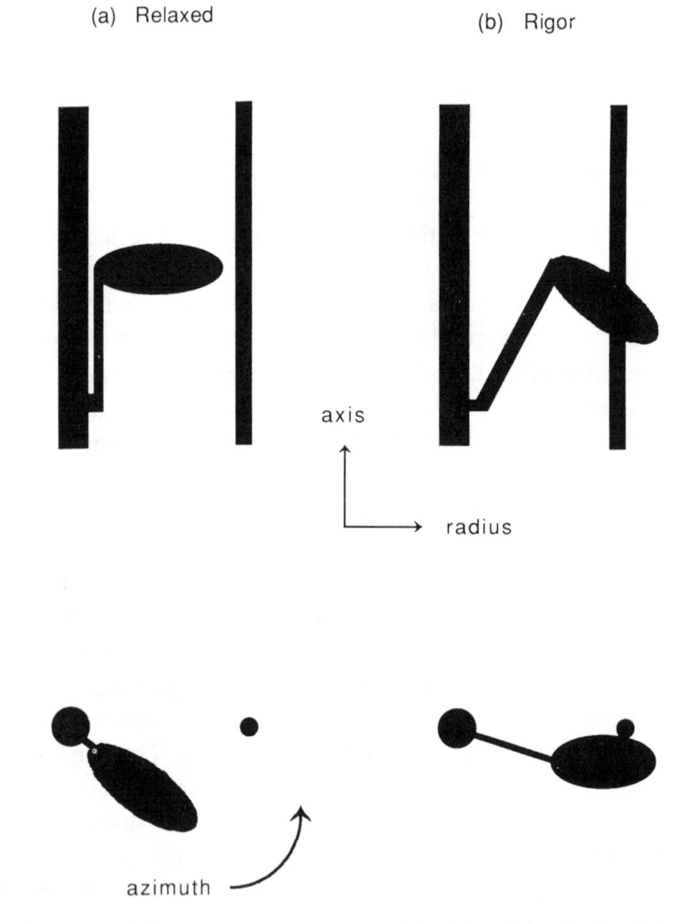

Fig. 6.8 Schematic diagram to show the potential axial, radial and azimuthal movements of a crossbridge. In contrast to many cartoons of the crossbridge cycle (e.g. Fig. 7.1), scale drawings show that rather little axial and radial movement of the crossbridge is required to attach to the actin filament because the myosin head length is comparable to the distance between the surfaces of the thick and thin filaments (10–18 nm, see Appendix).

lengths, the heads are able to achieve a stereospecific bonding with actin because of the flexibility in the S1–S2 and LMM–HMM link regions (section 4.3).

With the relaxed and rigor states characterized in outline, we can return to the question of what X-ray reflections reveal about longitudinal movements during contraction. As in rigor, an isometrically contracting muscle practically loses the 42.9 nm layer line, while the intensity and width of the 14.3 nm meridional may change slightly. Thus the crossbridges move from their helically ordered state about the thick filament backbone. A considerable effort has been directed towards detecting an increase in the actin layer lines to prove that the crossbridges actually attach to actin during contraction. The most serious charge we can bring using the equatorial data alone is one of loitering with intent. A significant increase has now been detected in the 5.9 nm spot [56]; however, in general, the results suggest that if a significant number of heads are attached then they do not take up a discrete angle, as in rigor. Changes are also seen in the 2nd and 3rd actin layer lines, far from the meridian, which are believed to report on the position of tropomyosin in the actin groove (section 8.2).

6.4 SPECTROSCOPIC STUDIES

As alluded to in section 4.3, fluorescent and spin-label probes have been attached to the myosin to report on rotational dynamics of the head. These same probes can also be used to label myosin heads within glycerinated fibres, with reasonable specificity for the reactive SH1 cysteine residue [see Thomas, 1987]. By aligning the labelled fibre relative to the plane of polarization of the excitation light or the magnetic field, information about the orientation of the heads may be deduced. Although the same SH1 group is labelled, the dipoles of different probes are not oriented the same way relative to the head and therefore they can reveal different results. Nevertheless, the general conclusions from these studies are that in relaxed muscle the probes, and hence presumably the heads, are highly mobile and do not show a strongly preferred angle [57]. This is in contrast to the X-ray results and some electron microscopy studies which indicate that the heads project at regular intervals and take on the helical symmetry of the thick filament. There are two possible explanations. Firstly, it is known that modification of the SH1 group does interfere with the ATPase properties of the myosin and, in particular, its ability to be fully inhibited by the regulatory system in the absence of Ca^{2+} (section 8.2). Thus the myosin heads may not be able to achieve a truly relaxed state (as opposed to a detached state) once labelled with the probe. Alternatively, it is possible that the SH1 cysteine residue is located in a locally flexible region so that the probe may take on different orientations (cf. section 4.3). However, when combined with the thin filament in the rigor state, the probes become rigid and highly ordered, in agreement with X-ray and electron microscopy data.

The extremely well ordered spin-label probes observed in rigor muscle indicate there must be considerable elasticity somewhere in the region between the site of labelling (SH1, see section 4.4) and the thick filament shaft. This is necessary to account for a constant angle of attachment of the myosin head to actin, despite the different periodicities of the two filaments. Furthermore, applying tension to a muscle in rigor does not perturb the narrow angular distribution of probes (Figure 6.9). A similar conclusion was reached when a fluorescent nucleotide diphosphate was bound to the myosin active site and tension similarly applied. These studies emphasize the important properties of the neck region of the myosin (20K domain), which may be able to bend relative to the rest of the head [58].

Under conditions of isometric contraction, spin-label probes indicate most of the heads are disordered, with about 10–20% in a rigor-like state. While initial studies assumed that the latter reflected the total attached population, further work concluded that a considerable fraction of the disordered population was attached but mobile on the microsecond time scale. These results are therefore in general agreement with the X-ray data, but they also serve to emphasize the difficulty of actually pinning down the nature of the movements of attached heads [57].

6.5 MECHANICAL STUDIES

The independent force generator model implies that the crossbridges behave as members of a tug-of-war team. We can learn about the opposition by examining their response to rapid length and tension

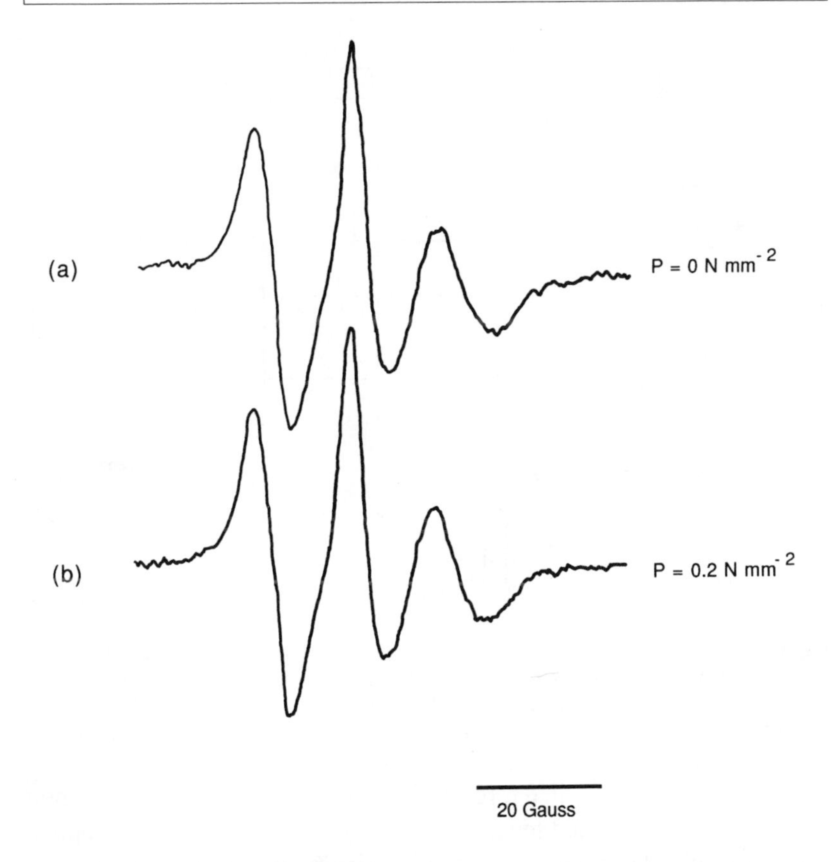

Fig. 6.9 ESR spectra of glycerinated rabbit psoas muscle fibres, labelled predominantly on the reactive SH1 of the myosin head. The fibres were aligned with the magnetic field. In the rigor state (a), the heads take on a limited range of angles with respect to the fibre axis, and hence the magnetic field, which results in a characteristic spectrum narrower than that of a randomly-distributed powder spectrum (cf. Fig. 4.10c). (b) Application of a force of 0.2 N mm^{-2} does not change this angular distribution. (From Cooke [58].)

changes. If we suddenly release the rope in a game of tug-of-war, it will move rapidly as the opposing team fall backward, then it will come to a halt as they pick themselves up and finally it will be withdrawn at a steady velocity as the team runs backwards (a so-called velocity transient). Alternatively, if we release the rope by only a small amount, the opposing force will drop instantaneously as the team are caught off guard, but will partially recover as they take up the slack by bending elbows and knees.

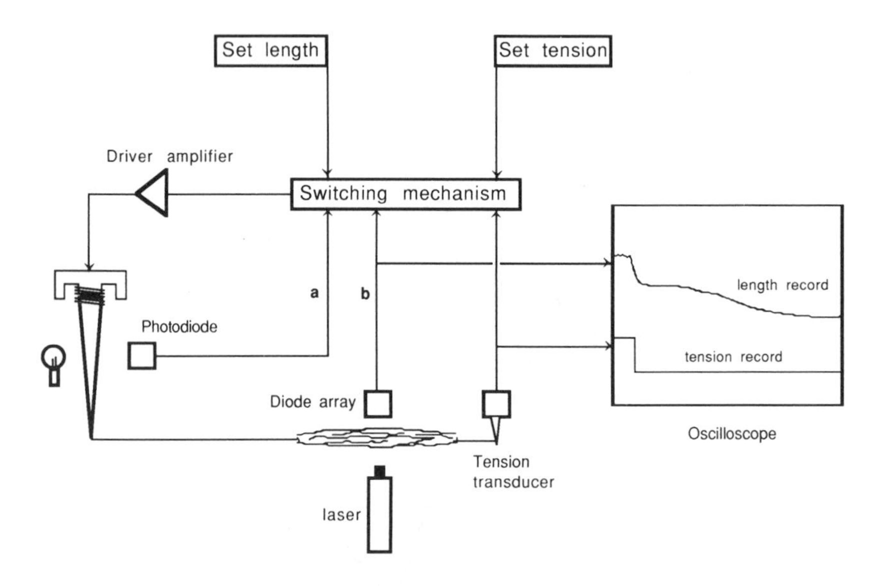

Fig. 6.10 The basic components of apparatus for recording the length and tension of a muscle fibre with high time resolution and servo control. Either the total length of the muscle is controlled via direct feedback from the servomotor (route a) or the sarcomere length of a central portion of the muscle is controlled via its optical diffraction pattern (route b).

Somewhat later the original force will be restored as they reoptimize their foot holding (a so-called tension transient). Without seeing the opposition we could determine the effective stride and reaction time of an individual, but we could not tell whether the arms, torso or legs were used to take up the slack.

Contracting muscle yields a similar response to such perturbations, although it is difficult to make the applied step changes rapidly enough to take the cross-bridge by surprise [47]. Mechanical devices designed for whole muscles (Figure 2.3) have a high inertia and are not suitable for these studies. Figure 6.10 illustrates the basic layout of a system capable of measuring the steady-state and transient behaviour of single fibres, just a few millimetres long with a response time below 1 ms.

A tension transducer converts the small distortion experienced by its stylus into an electrical signal. Such a device may rely on a change in its resistance or capacitance or may generate a current piezoelectrically. Either way, the physical movement of the stylus must be negligible relative to the desired length parameter imposed on the muscle. The length of the muscle can be controlled by a moving-coil device constructed by modification of a galvanometer or loudspeaker unit. Information about the length of the fibre can be fed back to the coil so that it performs as a servomotor. The method for monitoring the length can take various forms. Frequently it is based on the movement of a vane between a light source and a photocell. The signals from the length and tension transducers are amplified and recorded on an oscilloscope or are captured using a fast A/D converter. These signals are also fed back to the servomotor, via an electronic switching mechanism, allowing the muscle to be kept at a constant length or tension. Length and tension transients may be imposed electronically, under computer control. Stiffness is often recorded by monitoring the tension response when the muscle length is oscillated sinusoidally. The sensitivity, response time and stability depends on the components and damping circuits used and are optimized for the problem under study.

A major difficulty in any mechanical measurement is the mode of attachment of the specimen. Fibres may be tied with thread, clamped with aluminium foil or stuck using an acrylic glue. Nevertheless, even if the ends of the fibre are not seriously damaged and a good attachment is secured, muscle has a tendency to contract non-uniformly (section 6.1). To maintain a constant sarcomere length over the central portion of the fibre, reference markers, such as specks of gold leaf, may be positioned on the fibre or the optical diffraction pattern can be monitored (section 3.1). The servomotor is then controlled by negative feedback with this as a reference, rather than the total length of the muscle [48].

The response of frog muscle to rapid tension and length steps comprises four phases (Figure 6.11) which have been interpreted by A.F. Huxley and Simmons [59] as follows. Phase 1 occurs concomitantly with the applied step. In a tension transient, the instantaneous tension reached, T_1, depends linearly on the size and direction of the applied step, indicating that it involves an elastic element which behaves as a Hookean spring. Phase 2 represents the rearrangement of attached crossbridges. This leads to a partial recovery of the tension to a value, T_2, which is a non-linear function of the step length (Figure 6.12). Phase 3 is a plateau caused by detachment of strained crossbridges and reattachment at more optimally placed sites. In phase 4 reattachment continues until the original isometric tension, T_o is recovered or a steady isotonic contraction ensues.

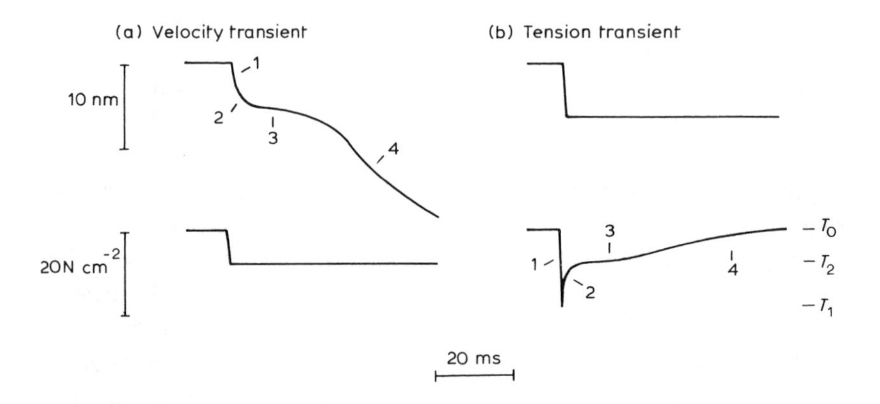

Fig. 6.11 The response of a frog skeletal muscle in isometric tetanic contraction at 2° C to (a) a rapid tension change, and (b) a rapid length change. The upper traces are the length records and the lower traces are the tension records as a function of time. The four phases of the transients are described in the text. (After A.F. Huxley [47].)

The amplitudes of the transient phases (T_1 and T_2) become scaled down at long sarcomere lengths (i.e. reduced filament overlap) in proportion to the isometric tension. Therefore the observed responses are attributed to the crossbridges rather than some series elasticity that might arise from components such as titin. The interpretation that phases 1 and 2 represent

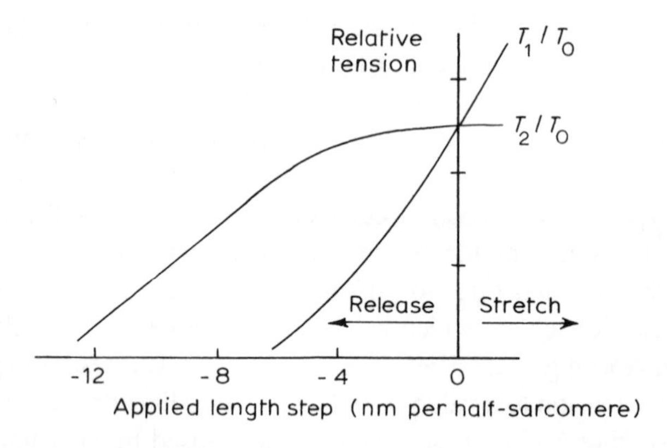

Fig. 6.12 The dependence of the instantaneous tension, T_1, and the early recovery, T_2, observed in a tension transient (see Fig. 6.11b) as a function of the size of the applied stretch or release of a muscle fibre in isometric contraction (T_o). (From A.F. Huxley [47].)

rearrangement of attached crossbridges, rather than detachment and reattachment, is less certain. If the linear dependence of T_1 on the step change is taken as evidence for passive elasticity in the crossbridge (Figure 6.12) then its slope (i.e. stiffness) provides a measure of the number of crossbridges attached. If a second length step is applied at the end of phase 2, the stiffness remains the same, suggesting that there is no change in the number of attached crossbridges at this point in time.

Assuming the Huxley–Simmons interpretation is correct, Figure 6.12 provides a measure of the working stroke of the crossbridge. Note that the applied length steps are expressed in terms of nanometres per half-sarcomere, the effective unit of contraction. The T_2 curve shows that the attached crossbridge can take up about 5 nm of slack to restore the isometric tension. Since, on average, the passive elastic element is effectively stretched by 6 nm from its rest length (the T_1 curve) in the isometric state, the total movement of the crossbridge is about 11 nm. This is in accord with the movement produced if the myosin head (15 nm) were to swing from an angle of 90° to 45° while attached, although obviously it does not prove that this is the precise mechanism.

Another notable feature of phase 2 is that its rate constant is larger (about 1000 s^{-1}) for greater releases. Huxley and Simmons [59] explained this on the basis of several attached states. Under isometric conditions, these are in a balanced equilibrium because chemical energy is traded for mechanical energy (extension of the elastic element). A quick release will perturb this distribution: the greater the release, the less the strain in the elastic element and so the lower the activation energy for redistribution. This model differs from the Huxley 1957 scheme [46] in several respects (cf. section 6.1). There, crossbridges were proposed to attach in a strained state, but this is undesirable on kinetic grounds because it requires that thermal motion overcomes an activation energy of the same magnitude as the overall transduction process. By postulating several attached states, the activation energies for the individual steps, including initial attachment, are reduced. However, the original postulate that the rate of detachment for crossbridges exerting positive tension is slow, remains important for efficient coupling.

As with the tug-of-war game, the domains within the crossbridge responsible for these transients cannot be identified without further information. A number of possibilities have been considered for the location of the elastic element, including domains within actin. Many diagrams show a spring-like component within S2, although the grounds for this are usually no more than the ease of drawing! In this location S2 itself must be capable of extending about 11 nm. Melting studies of long S2 and intact myosin have shown that about 150 residues in the region of the LMM–HMM junction are particularly labile. A helix–coil transition in S2 could account for the length changes of the magnitude observed in the mechanical transients. However, the untwisting of a coiled-coil helix would cause topological problems for a myosin molecule in a filament, particularly if the heads are to remain attached to the actin. The other extreme is to consider that the elasticity is in the actomyosin bond itself. This could undergo angular distortion and allow the head to achieve 11 nm movement at its distal end. However, spectroscopic studies do not favour this option (section 6.4). A compromise hypothesis involves an elastic bending of the neck region of S1.

6.6 COMBINED STUDIES

The techniques described above give a complementary, and sometimes a conflicting, view of crossbridge structure and function. In order to clarify and extend these investigations, considerable effort has been directed towards their simultaneous application. Now that the X-ray diffraction pattern can be recorded on the millisecond timescale, it has been possible to combine these studies with mechanical perturbations [60]. As noted in section 6.3, the 14.3 nm spot, arising from the axial spacing of the crossbridges, is detectable in relaxed, rigor and isometrically contracting muscle, indicating that the crossbridges can select actin sites which do not drastically perturb their regular origin along the thick filament axis. Following a rapid stretch or release, the intensity of the 14.3 nm spot drops markedly, but recovers within about 50 ms. If a quick stretch is applied soon after a quick release, or *vice versa*, some of the intensity is regained immediately [60]. These results imply that, following a length perturbation, many attached crossbridges are distorted longitudinally so that their mass is no longer distributed at 14.3 nm intervals. Subsequently, the crossbridges detach and find more favourable actin sites.

Isotonically contracting muscle is more difficult to study since there is only a small time window during which data can be collected. Nevertheless, it has been possible to follow the X-ray pattern during rapid shortening (i.e. at velocities approaching V_o). During this phase, the 14.3 nm spot remains at low intensity, possible due to compression of the crossbridges before they detach. The equatorial reflection ratio, 1.1/1.0 falls during shortening, in accord with the idea that fewer bridges are attached. However, the kinetics of this process lag behind the fall in tension following the quick release, suggesting that the crossbridges cycle several times before coming to their new steady-state distribution. In other words, crossbridges which are detached during active contraction are not in the same state as those in relaxed muscle [54].

Following the observation that the filaments in asynchronous, insect flight muscles (section 2.1) are well ordered in the electron microscope (Figure 6.2), these muscles have provided an important source for characterizing crossbridge structure and function [see Poole *et al.*, 1988]. It is therefore pertinent to consider some of their unique properties. The I-bands are very short and the thick filaments are connected to the Z-line by an elastic protein, presumably analogous to titin, which gives it a high resting stiffness. When stimulated, the tension increase is delayed until the fibre is slightly stretched (so-called stretch activation). In the insect this property allows the flight muscle to drive an inertial load (i.e. the wings plus ligaments) at its natural frequency of oscillation, in the same way as a child on a swing is kept in motion. This phenomenon also accounts for the asynchrony between the nerve input and the oscillatory contractions of the muscle which are self-sustaining [see Rüegg, 1986, Chapter 8]. The muscle only shortens by a few percent of its rest length, which is the same order of filament sliding that may be achieved by a single crossbridge cycle. Stretch activation provided a useful experimental tool for synchronizing time resolved X-ray measurements [see Poole *et al.*, 1988]. More recently contraction has also been synchronized using the release of caged-ATP (section 7.4). Note that the thick and thin filaments are arranged on a different hexagonal lattice to that of vertebrate skeletal muscle. The thin filaments are located between two thick filaments, in contrast to Figure 6.4, and give rise to different equatorial reflections; the 1.0 plane samples actin and myosin, while the 2.0 plane samples actin alone. The equatorial reflections show large changes on going from the relaxed to rigor state, as observed for vertebrate muscle. In contrast, however, the equatorial ratio for the tension bearing state following a stretch (presumably equivalent to the isometric state) does not show a large shift from the resting value, although the 14.5 nm spot, corresponding to the axial repeat of the thick filament, does show a large drop. Stiffness measurements indicate a significant proportion of the cross-bridges are attached following stretch activation, and therefore presumably they take up a different position to those in rigor [see Poole *et al.*, 1988].

With this background to the dynamics of crossbridges in contracting muscle we now consider how their movements might be coupled to ATP hydrolysis.

TOPICS FOR FURTHER READING

Ingels, N.B.(ed.) (1979) *The Molecular Basis of Force Development in Muscle*, Palo Alto Research Foundation, California. (Contains a useful summary of contraction theories postulated prior to the 1980s.)

Huxley, A. (1980) *Reflections on Muscle*, Liverpool University Press, Liverpool. (A penetrating historical view of the development of crossbridge theories.)

Squire, J.M. (1981) *The Structural Basis of Muscular Contraction*, Plenum Press, New York. (A comprehensive treatise on structural aspects of muscle, with a good introduction to diffraction theory.)

Thomas, D.D. (1987) Spectroscopic probes of muscle crossbridge rotation. *Annual Review of Physiology*, **49**, 691–709. (Summarizes results from optical and ESR spectroscopy.)

Sugi, H. and Pollack, G.H. (eds) (1988) *Molecular Mechanism of Muscle Contraction*. Advances in Experimental Medicine and Biology, **226**, Plenum Press, New York, London. (Comprises papers from a recent symposium, complete with lively discussions.)

Poole, K.J.V., Rapp, G., Maeda, Y. and Goody, R.S. (1988) Synchrotron radiation studies on insect flight muscle. *Topics in Current Chemistry*, **147**, 1–29. (Covers technological aspects of X-ray diffraction methods.)

Wakabayashi, K. and Amemiya, Y. (1991) Progress in X-ray synchrotron diffraction studies of muscle contraction, in *Handbook of Synchrotron Radiation*, Vol. 4, (eds S. Ebashi, M. Koch and E. Rubenstein), Elsevier, Amsterdam. (A comprehensive review of recent X-ray studies.)

Mechanochemical coupling 7

7.1 BIOCHEMICAL SCHEMES

The Lymn–Taylor scheme, outlined in section 5.3, provides a simplistic mechanism for mechanochemical coupling in which the alternation between the actin-associated and actin-dissociated intermediates (equation 5.7) corresponds, in a one-to-one fashion, with the alternation between attached and detached states of the crossbridge [42]. Lymn and Taylor specifically identified detachment with the A.M.T → M.T step and attachment with the M.D.P → A.M.D.P step (in this, and subsequent models ATP, ADP and Pi are abbreviated to T, D and P respectively). Relative movement of the actin and myosin filaments was proposed to occur by a swinging crossbridge mechanism (90° → 45° angle change) coupled to the products release step A.M.D.P → A.M, which geometric considerations indicate would achieve about a 10 nm step size. This mechanism (Figure 7.1) provides the start point of most discussions of mechanochemical coupling.

Apart from its historical significance, the scheme is useful to illustrate the concepts involved in a more quantitative description of mechanochemical coupling (section 7.2). However, its limitations must also be considered. As discussed in section 5.3, subsequent studies of the actin-activated S1 ATPase in solution showed that M.T and M.D.P had comparable affinities for actin, and that the hydrolysis step could occur while the head remained attached to actin (i.e. A.M.T → A.M.D.P). Thus the ATPase cycle need not necessarily be tightly coupled to the mechanics. Nevertheless, the essential concepts of the scheme could be retained by considering that myosin intermediates alternate between weak bind-

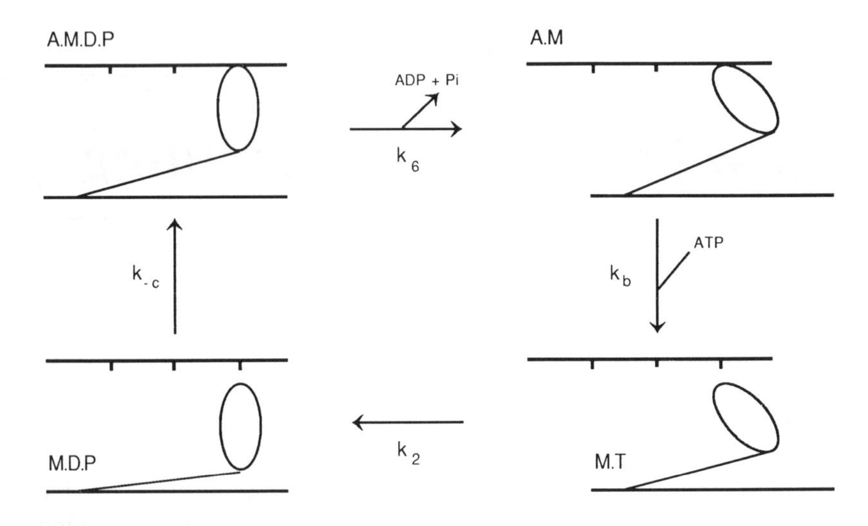

Fig. 7.1 The Lymn–Taylor model for the correlation between the chemical and mechanical states during the crossbridge cycle [42].

ing states (M.T and M.D.P) and strong binding states (M.D and M). In this model the global change in S1 structure, equivalent to a swinging crossbridge, was assumed to be associated with the transition from the weak- to strong–binding states.

Before further discussion of these models it is pertinent to ask whether it is quantitatively viable to link one ATPase cycle with one crossbridge stroke. Consider a frog muscle contracting at 0°C at optimal efficiency, i.e. lifting a load, $P_o/3$ at a velocity, $V_o/3$ (Figures 2.4 and 2.7). The thermodynamic efficiency of contraction is defined by the ratio of the power output (work rate) to ATP utilization rate. From the data in the Appendix, the power from 1 cm^3 of muscle is given by:

$$P_o/3 \times V_o/3 = 6.7 \times 0.67 \text{ N cm s}^{-1} = 0.044 \text{ J s}^{-1}$$

Assuming 1 cm^3 = 1 g of muscle and -ΔG for ATP hydrolysis = 60 kJ mol^{-1}

$$\text{ATPase rate} = 1.5\ \mu\text{mole s}^{-1} = 0.09 \text{ J s}^{-1}$$

Therefore the thermodynamic efficiency of coupling = 0.044/0.09 = 0.49.

We may calculate the stroke size, or at least a limit, of the putative crossbridge cycle by two routes (these are not independent).

1. From the work done (force × distance) per ATP molecule.
   Let a fraction F of the 300 myosin heads in half a thick filament be attached and move, on average, through a distance a nm. Then:

   Force = $P_o/3 = 6.7\ \text{N cm}^{-2} = 1.1 \times 10^{-10}$ N per filament
   $= 3.7 \times 10^{-13}/F$ N per head.

   Work done = $3.7 \times 10^{-13} \times a \times 10^{-9}/\text{F}$ J per head per cycle

   Energy available from 1 ATP molecule = $60 \times 10^3/6 \times 10^{23} = 10^{-19}$ J. Hence for an efficiency of 0.49, the crossbridge stroke, $a = 10^{-19} \times 0.49 \times F/3.7 \times 10^{-22} = 130F$ nm.

2. From the ATPase rate (6.3 ATP per head per s):
   For a unit of a half-sarcomere, $V_o/3 = 833\ \text{nm s}^{-1}$.
   During 1 ATPase cycle let the head actively move on average a nm while attached and free-wheel b nm while detached.
   Hence, $6.3(a + b) = 833\ \text{nm s}^{-1}$.
   For a large population of crossbridges, the fraction attached at any one instant, $F = a/(a + b)$, which is a measure of the duty cycle of the motor. Therefore, $a = 833F/6.3 = 130F$ nm.

While the actual stroke size, a, cannot be determined without independent knowledge of F, the calculations show that the 11 nm movement, deduced from mechanical studies (section 6.5) could correspond to the turnover of a single molecule of ATP without violating thermodynamics. Taking 11 nm as the average stroke size implies that F is rather small (8%) during isotonic shortening and indeed would fall to about 3% at V_o. For a large assemblage of crossbridges in a muscle this is not a problem, but it does raise a significant question *in vitro* when dealing with a small number of myosin heads (section 10.1).

These calculations predict that the fraction of heads attached should decrease as the speed of shortening increases, because a has an upper limit (about 20 nm) set by the dimensions of the crossbridge and the ATPase rate falls slightly at velocities above $V_o/3$ (therefore b must increase). Experimental methods for determining F are ambiguous. The 1.1/1.0 ratio of the equatorial X-ray reflections falls slightly at high velocities of contraction, but this number also depends on the angular distribution of the heads. Furthermore, this ratio does not distinguish between attached heads and those just near to the actin filament. The instantaneous stiffness (slope of T_1 in Figure 6.12) of an isotonically contracting muscle is 20% that of rigor, while the isometric state is about 70% as stiff as rigor [62,63]. Thus under isometric conditions each attached head can generate a tension of about $(3 \times 3.7 \times 10^{-13})/0.7 = 1.6$ pN. These findings are in qualitative agreement with the above calculation. Nevertheless, since all the heads are believed to be attached in rigor and if it is assumed that active attached states have a similar stiffness, then the observed value of F, in the isotonic case, is higher than expected from the calculations above. It is possible that a substantial number of heads attach transiently, thereby increasing F, without going through a complete ATPase cycle. This property complicates the measurement of the crossbridge stroke size from mechanical measurements of muscle fibres [64].

Note that the rate constants for processes occurring while the head is attached must exceed $833/a = 75$ s^{-1} (and hence > 225 s^{-1} at V_o). How do these data compare with solution kinetics? The V_{max} for the actin-activated S1 ATPase is about 4.5 s^{-1} at 0°C. As discussed previously (section 5.3), the V_{max} in solution obtained by extrapolation to infinite [actin] does not appear to measure a rate-limiting step between attached states. Rather, V_{max} is a composite function also involving dissociated states. For example, if V_{max} is limited by a refractory state transition in the dissociated head, then a better comparison of V_{max} with events in muscle would be given by the parameter $833/b = 7$s^{-1}. Thus the original correlation seen between V_{max} for the ATPase in solution in different muscle types and V_o for the maximum shortening velocity is not a direct one. It appears that all the elementary steps in a slow muscle are slower relative to a fast muscle to give this approximate correlation. A better match between the minimum rate constant for detachment of a crossbridge, calculated at V_o (i.e. 225 s^{-1} in the above example) has been noted with the dissociation rate of ADP from the A.M.D state [65]. Nevertheless, even this comparison is oversimplistic. Although the ATPase of isolated S1 heads might be expected to be more similar to a muscle contracting under zero load than to one contracting isometrically, the filament lattice still imposes structural constraints. There is no simple solution to this problem. A generalized approach is considered in the next section.

7.2 THE FILAMENT LATTICE PROBLEM

Although crossbridges show a considerably elasticity, they cannot behave totally independent of each other in a muscle fibre. The availability of an actin site for a particular crossbridge will depend on the relative displacement of the thick and thin filaments. Moreover, once bound, the rate constant for any transition which alters the angle or effective length of the head will itself depend on the tension borne by the sarcomere. The distribution of chemical states in a muscle will therefore be different from that in solution. If we assume that the elastic element in the crossbridge behaves as a Hookean spring (i.e. $\Delta T \propto \Delta L$), then the energy required to stretch or compress it is proportional to $\Delta L^2/2$. Simplistically, we can use data obtained in solution to calculate the equilibrium constant (and hence chemical free energy) at zero strain and then modify this number by an amount appropriate to the elastic energy [see Hill, 1989]. In this way the energetics can be defined for each step as a function of the filament displacement.

Figure 7.2a shows the Lymn–Taylor model (Figure 7.1) that has been modified to include an elastic element. It is assumed that the A.M.D.P state interacts with actin at 90° while the A.M state is at 45° to the filament axis. The elasticity is modelled as a spring in S2 for convenience (cf. section 6.5). The relative longitudinal position of the filaments for the crossbridge under consideration, x, is arbitrarily taken as 0 nm when the A.M.D.P state is unstrained. Hence if the head is 15 nm long, the A.M state will be unstrained when x = 15.sin 45° = 11 nm. In solution at physiological [ADP] and [Pi], the A.M.D.P $\rightarrow$ A.M transition lies strongly to the right ($K = 10^6$), but in the model (Figure 7.2a), this step can only be achieved by stretching the spring (state b) or allowing the filaments to move (state c). Hence the equilibrium between states a, b and c will depend on x. On the other hand, the equilibrium between states d and e is unaffected by the value of x and therefore the solution value for M.T $\rightarrow$ M.D.P should apply directly.

In order to keep the comparison of free energy changes meaningful, each step must be in the form of a unimolecular transition, i.e. binding steps are normalized by taking an effective reactant concentration. The standard free energy change so adjusted is termed the basic free energy [see Hill, 1989]. Plausible basic free energies of the states depicted in Figure 7.2a are shown in Figure 7.2b as a function of x. M.T and M.D.P are independent of x and are separated by only 5.4 kJ mol^{-1} because this

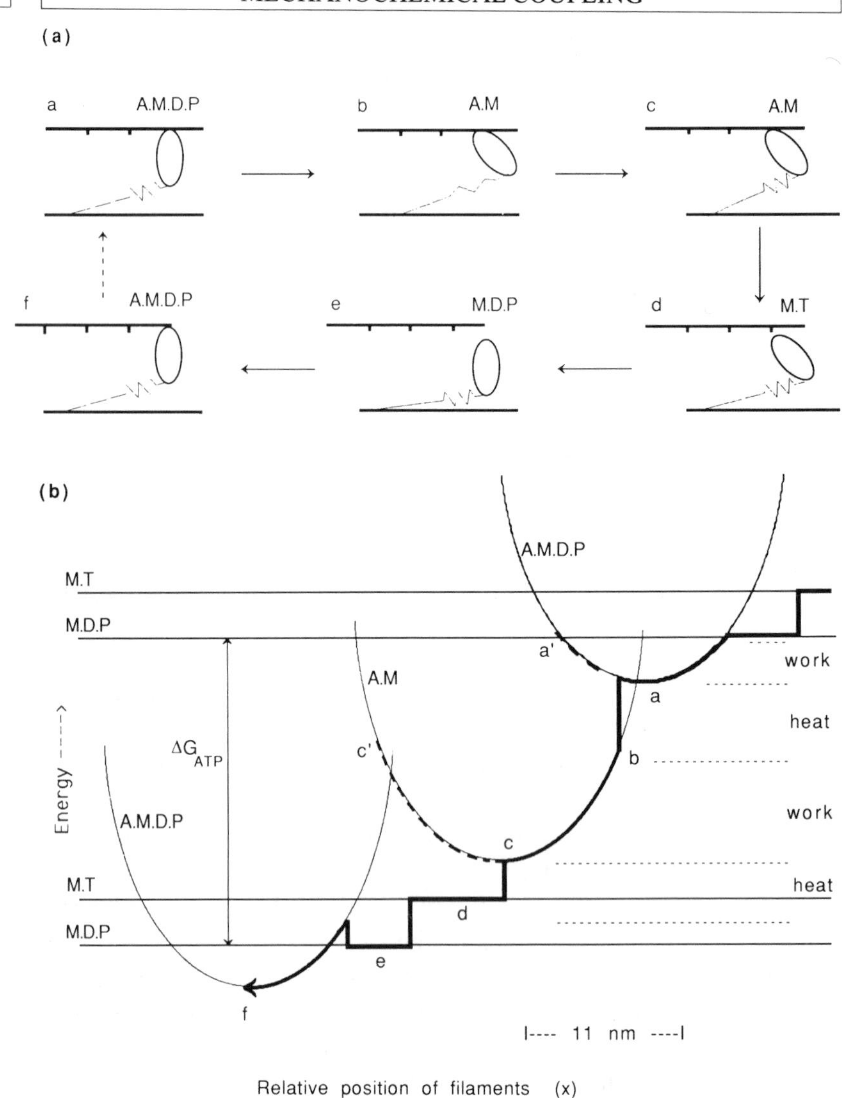

Fig. 7.2 (a) The Lymn–Taylor model modified to include an elastic element capable of maintaining tension. (b) The corresponding energy diagram for the various chemical states. The thick line represents one possible route for a crossbridge during filament sliding at moderate velocities. Note that work is only performed when the force from the elastic element operates over a distance. Vertical energy drops result in loss as heat. The dashed line represents a possible route at maximum shortening speed, V_o, when the elastic element may be compressed as the crossbridge is dragged through its minimum mechanical energy (see text).

reaction is readily reversible (equation 5.4; $K_2 = 9$). Attached states show a minimum energy defined by their basic free energy in solution and increase either side as a function of x^2 due to the energy of compressing or stretching the spring. The overall free energy drop for one cycle equals that for ATP hydrolysis (- 60 kJ mol^{-1}; section 2.5).

Provided that the energy minima of the two attached states are displaced with respect to x then work can be extracted from the cycle. The solid line in Figure 7.2b shows a possible path taken by a crossbridge which would give efficient coupling during isotonic contraction. Work is only obtained by releasing the elastic energy of a state as x changes (e.g. b $\rightarrow$ c). Free energy released during chemical transitions (shown as vertical drops) is lost as heat. No work can be extracted from transitions when the crossbridge is detached (e.g. d $\rightarrow$ e), hence the finding that in solution these reactions have only a moderate equilibrium constant is self-consistent. In an isometric contraction x is held constant (although crossbridges throughout the filament will have different x values due to the mismatch of the actin and myosin periodicities), hence only vertical transitions can occur. Isometric tension is generated by crossbridges fluctuating between states a and b, so trading chemical for mechanical energy, but all the free energy is ultimately released as heat.

When the muscle is shortening without an external load, it is possible that attached states will be drawn into regions where they develop negative tension. Thus an attached head in the A.M.D.P state may continue to position a′ (Figure 7.2b) and then dissociate back to M.D.P. No work is done and no chemical energy is lost, but such weakly attached states would generate a drag force. This situation would lead to a reduction in the overall ATPase rate at V_o, since fewer heads would have time to complete a cycle (cf. section 6.1). Strongly bound states will also be drawn into regions of negative tension development at high velocities of shortening (e.g. state c′), again reducing the net work output. It is possible such heads will be physically dissociated from actin, before they have time to bind ATP, giving the nucleotide-free M state (and also the M.D state, which is not shown in the simplified scheme in Figure 7.2). The properties of these torn heads have yet to be defined. If the mechanical energy store were simply an elastic element, as shown in Figure 7.2a, then it would be expected that this energy would be lost. However, physiological evidence has indicated that such heads might rapidly recombine with actin units further along the filament and complete their cycle [66].

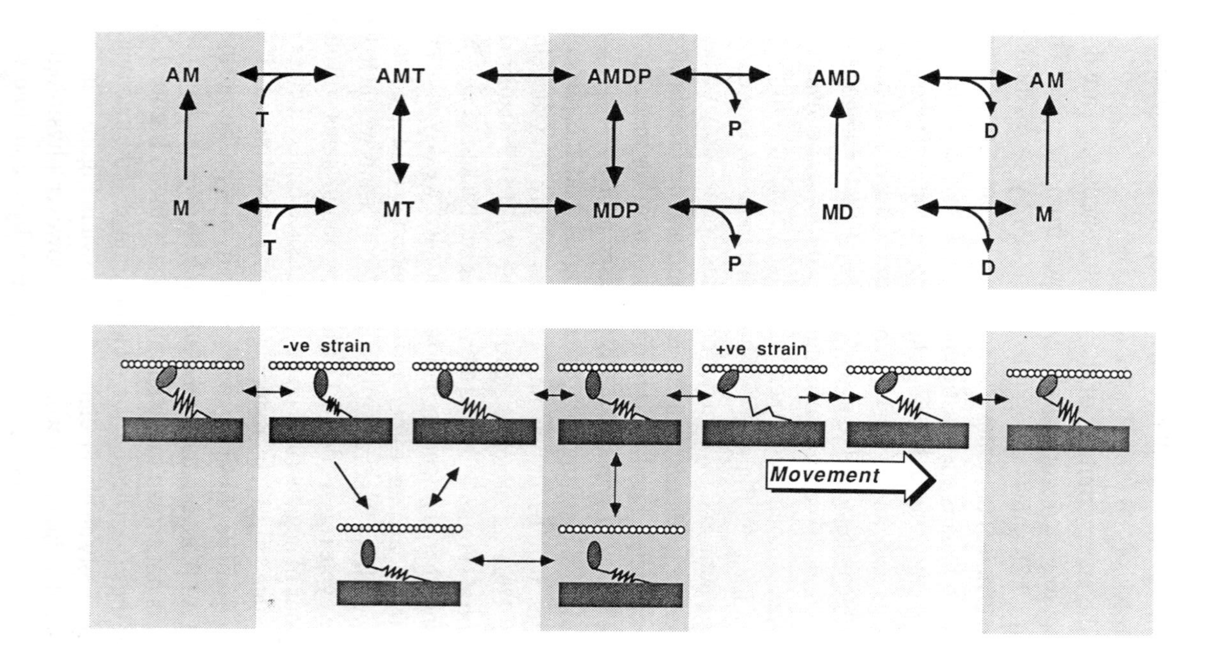

Fig. 7.3 A crossbridge model in which the chemical states correspond to a weak-binding (90°) state or a strong-binding (45°) state. The A.M.T and A.M.D.P states) are roughly isoenergetic and no active 're-cocking' of the head is necessary for the crossbridge when dissociated (cf. Figs 7.1 and 7.2). The reversal of the crossbridge stroke occurs during the very short-lived negatively strained A.M.T state [67,68]. (Courtesy of Dr N.C. Millar.)

Figure 7.2b is not a solution for the mechanism of contraction but rather one way of formulating the problem. One crucial assumption is that the energies of the A.M.D.P and A.M states have their minima at different x values. The other crucial assumptions for mechanochemical coupling concern the rates of the transitions. In Figure 7.2b the energy levels define the equilibrium constants and hence they only fix the ratio of the forward and reverse rate constants and not their absolute magnitude. We must assume that a direct b $\rightarrow$ d transition is prevented (i.e. it is very slow), otherwise state c will be bypassed and little tension or work could be obtained (equivalent to Huxley's assumption about g [46]; section 6.1). Overall this short circuit is energetically favourable and therefore must be blocked by a high activation energy, in the same way as the spontaneous hydrolysis of ATP is prevented. In the absence of detailed structural information about S1 and the actin binding site, an explanation for this valve-like property cannot yet be formulated.

This kind of analysis has also been applied to more detailed biochemical schemes in which the Pi and ADP release steps are kinetically resolved and the weak binding properties of the M.T and M.D.P states are considered [see Brenner, 1990]. Experimental evidence suggests that Pi is released first and the structural properties of the resultant A.M.D complex are similar to the A.M state. Thus in Fig 7.2b the A.M.D free energy profile would lie close to the A.M curve and would represent the major force generating state. For reasons of simplicity the weak binding and strong binding states have been considered to bind preferentially at 90° and 45° respectively (Figure 7.3). This assumption leads to a reversal of the crossbridge stroke during the A.M $\rightarrow$ A.M.T step prior to dissociation [67]. However, because the A.M.T state is short lived, the negative tension developed is outweighed by the other (more numerous) crossbridges in the filament in the force generating A.M.D state [68]. If this were an exclusive route, however, it implies that a single head could not support the net movement of the thin filament (cf. section 10.1).

7.3 CHEMICAL STATES IN MUSCLE

From the above considerations we would expect some similarities and some differences in the kinetics of chemical transitions in solution compared to those in a muscle fibre. Indeed the steps that differ are, by implication, the ones that are involved in the tension-generating process. Several approaches have been used to correlate the chemical states observed in solution with those which exist in a muscle fibre. The magnitude of chemical and mechanical rate constants themselves provide a crude method of comparison. More critical assessments can be achieved by measuring the effect of perturbing the chemical composition of permeabilized fibres on the tension and shortening speed, either in the steady-state or transiently [see Goldman, 1987; Homsher and Millar, 1990]. Chemical states in muscle can also be explored by following the fate of labelled substrates and products. Let us start with some semi-quantitative comparisons.

Solution studies have categorized actomyosin intermediates into two classes; weak-binding and strong-binding states. Do they have counterparts in muscle? Weak-binding states correspond to the M.T and M.D.P intermediates. In solution, the observed actin binding constants for these states are of the order of 100 μM, but the affinities are increased at low ionic strength. To explore for the presence of weak-binding states in muscle fibres, stiffness measurements were carried out at low ionic strength in the presence of ATP and in the absence of Ca^{2+}. It was found that the observed stiffness was dependent on the speed of stretch and thus attached crossbridges were detaching during the measurement. Even the fastest stretches (complete in 0.15 ms) were not sufficient to completely freeze the dissociation process [see Brenner, 1990; 69]. The estimated dissociation rate of these weakly attached crossbridges was about 10^4 s^{-1}, comparable to solution studies (e.g. k_b of equations 5.7 and 5.8). By comparison with rigor stiffness, it was concluded that up to 50% of the heads may be bound under these conditions. At physiological ionic strengths, some weakly attached crossbridges were detected by rapid stretches, although their population was much smaller. The ratio of the 1.1/1.0 equatorial X-ray diffraction spots, as a function of ionic strength, supported the conclusion that crossbridges can attach under relaxing conditions. Moreover, the absolute intensities of the equatorial reflections, and also other reflections, indicated that the attached heads were not in the same configuration as those of rigor [see Brenner, 1990].

Strong-binding states are expected to predominate in the absence of nucleotide (rigor) or in the presence of ADP. Stiffness measurements indicate that crossbridges detach on the time scale 10^{-2} to 10 s^{-1}, i.e. several orders of magnitude slower than for weakly attached crossbridges. In the presence of ATP under activating conditions, a mixture of weakly and strongly attached states is expected. By using moderate speeds of stretch

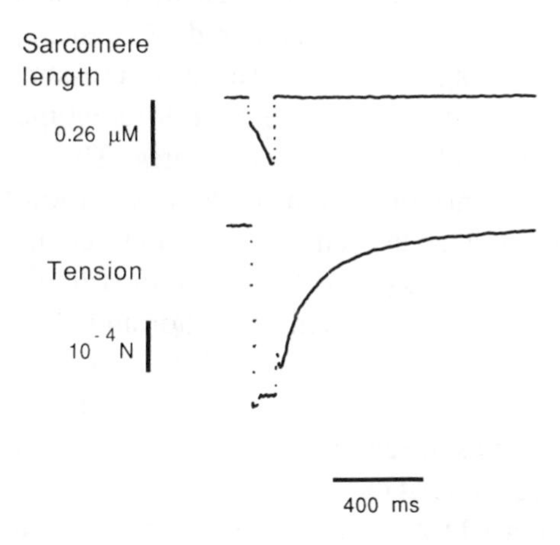

Fig. 7.4 The kinetics for force redevelopment when a rabbit psoas muscle, undergoing a brief isotonic shortening is restored to its original length. The upper trace represents the sarcomere length and the lower trace the tension response. (After Brenner and Eisenberg [49].)

and moderate ionic strengths (0.17 M) the contribution from weakly attached states can be minimized. The remaining crossbridges appear to dissociate with rate constants in the order of 50 to 1000 s^{-1}.

At physiological ionic strengths, the weakly attached intermediates are largely in the dissociated state and hence do not contribute significantly to the tension developed by the fibre. Force development is therefore associated with the transition from the weakly attached to strongly attached states. This process also controls the flux through the ATPase cycle of isolated acto–S1 (section 5.3). How do these rates compare experimentally? Brenner and Eisenberg [49] compared the rate of ATPase activity in solution with the rate of tension recovery when an unloaded isotonically contracting muscle was rapidly stretched back to its starting length (Figure 7.4). Cross-linked acto–S1 was used for the solution measurements to avoid the problem of the large extrapolation to infinite [actin], necessary at moderate ionic strengths. The mechanical protocol was chosen because during an unloaded shortening most (80%) of the heads are detached in rabbit psoas muscle and are in a steady-state distribution. The recovery of tension (around 20 s^{-1} at 20°C) roughly matched the ATPase rate over a 30°C temperature range. Moreover, they are in accord with the values of f, the effective crossbridge attachment rate constant, that A. F. Huxley [46] used in his 1957 model (section 6.1). The expected rate constant for tension redevelopment is actually given by $f + g$, which approximates to f because for crossbridges developing tension, $g < f$. Note that f and g take on a different meaning in models with more than one class of attached states. They are the effective composite rate constants for the transitions between non-force generating and force generating states, and are not the actual association and dissociation rate constants of weakly bound or strongly bound intermediates. A more detailed comparison of solution and fibre kinetics requires consideration of the lattice problem, as alluded to in section 7.2.

Analysis of the effect of changing the concentration of ATP, ADP and Pi on the tension and shortening velocity of muscle fibres in the steady-state is a direct analogue of the steady-state treatment of enzyme kinetics in solution (section 5.1). According to in Figure 7.2, increasing the [ATP] will speed up the c $\rightarrow$ d transition and hence increase the overall cycling rate. In practice, the shortening velocity of muscle shows an approximately hyperbolic dependence on [ATP], but the apparent K_m is an order of magnitude higher than that observed for the actin-activated ATPase activity in solution under similar ionic conditions [70]. This may be accounted for by the effect of comparatively few rigor (A.M) crossbridges, present at intermediate [ATP], which would give a disproportionate resistance to motion produced by the cycling crossbridges that spend most of their time in the dissociated state. In contrast, tension is reduced as the [ATP] is increased to a physiological value, because the A.M state is capable of bearing force (cf. Figure 7.2).

Addition of Pi or ADP to muscle in rigor has very little effect. In the former case it is unlikely that Pi binds to the A.M state to any significant extent. In contrast, ADP does bind to A.M to give an A.M.D state although the mechanical and structural properties are little changed from rigor (hence the assignment of similar structural states in Figure 7.3). More revealing are the effects of these products when added in conjunction with ATP [70].

Pi is a very weak inhibitor of the actomyosin ATPase in solution. On the other hand, millimolar concentrations of Pi significantly depress the isometric tension generated by a muscle in the presence of ATP, as well as reducing its stiffness [see Goldman, 1987; Homsher and Millar, 1990]. These effects are at least partly due to Pi binding to the active site, as indicated by the extensive ^{18}O exchange observed between H_2O and the added Pi (so-called medium exchange). It seems that Pi binds more readily to an A.M.D state in a muscle fibre compared with that in solution, and drives the kinetic scheme (equation 5.8 and Figure 7.3) backwards to give weakly bound states (e.g. A.M.D.P). Once Pi is bound, a small amount of A.M.ATP and M.ATP will form, thus accounting for the medium ^{18}O exchange (cf. section 5.2). Medium exchange is more extensive in isometrically contracting muscle than in relaxed muscle [71]. Hence tension appears to perturb the distribution of chemical states to favour an A.M.D state that is not highly populated in solution. In turn, this argues that the Pi release step is associated with a transition in muscle which produces a major force generating state of the crossbridge.

Pi has little effect on the maximum shortening velocity, V_o. This can be accounted for in terms of Figure 7.3 since movement is produced by the release of tension from the A.M.D (or A.M) state, a process which is independent of Pi. Increasing Pi will reduce the number of crossbridges which participate in sliding but this will not affect the net velocity since very few crossbridges are required to operate when the fibre is unloaded. The loss in tension on increasing Pi can also be seen in terms of a reduction of the free energy available from ATP hydrolysis (i.e. an increase in Q in equation 2.3). This would reduce the vertical distance between states a and c in Figure 7.2b, giving a reduced force, but it would not effect their horizontal separation which controls the translation of filaments. During fatigue the [Pi] increases *in vivo*, as determined by NMR methods (Figure 2.6), and may account for the drop in performance of a muscle [1]. After carrying a heavy suitcase it is relatively easy to flex one's muscles when unloaded but it is difficult to lift another weight.

ADP has an opposite effect to Pi. The isometric tension generated in the presence of ATP is increased by ADP addition, while the maximum velocity of shortening is reduced [70]. These effects can be explained with reference to Figure 7.3 where an increase in ADP will favour the tension-generating A.M.D state, but it will inhibit crossbridge cycling by competing with ATP for the A.M state. To a rough approximation, increasing [ADP] at constant [ATP] has a similar effect to reducing [ATP], indicating that the A.M.D and A.M states have similar minima for their free energy profiles (cf. section 7.2).

7.4 CHEMICAL KINETICS IN MUSCLE

As described in Chapter 5, rapid mixing methods have played a key role in identifying intermediates of the actomyosin ATPase. Unfortunately, rapid mixing cannot be used to resolve reactions on the millisecond timescale in muscle fibres because events are limited by the diffusion time of nucleotide. This problem has been overcome by the use of caged compounds [see Goldman, 1987; Homsher and Millar, 1990]. These are inert, photolabile derivatives of substrates or ligands which can be diffused into a muscle fibre or other cells and rendered active by a brief flash of intense light. Photolysis of caged-ATP (Figure 7.5) yields millimolar concentrations of ATP in a few milliseconds [72].

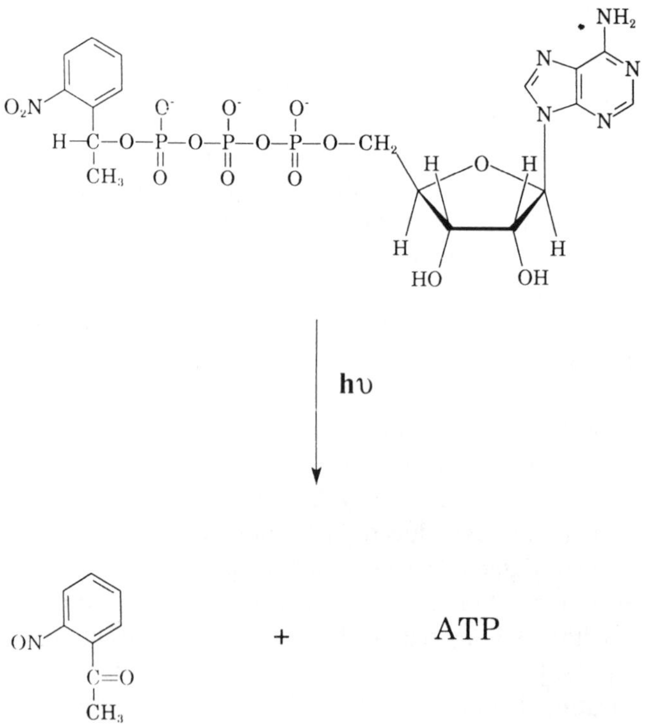

O_2N H—C—O—P—O—P—O—P—O—CH_2 CH_3 O O O O^- O^- O^- NH_2 N N N N H O H H H HO OH

hυ

ON C=O CH_3 + ATP

Fig. 7.5 Photolysis of P^3–1–(2-nitro) phenylethyladenosine 5′-triphosphate (caged–ATP) to generate ATP.

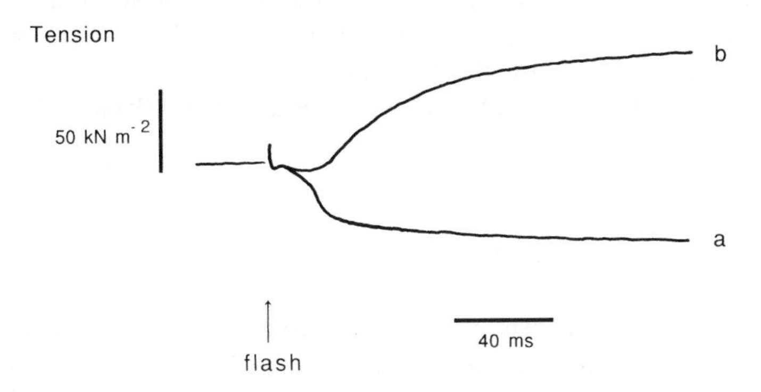

Fig. 7.6 Tension transients initiated by the photolysis of caged-ATP (↑) in rabbit psoas muscle fibres, initially in rigor, in (a) the absence of Ca^{2+} and (b) in the presence of Ca^{2+} (After Goldman *et al.* [72].)

In a muscle initially in rigor, photolysis of caged-ATP, in the absence of Ca^{2+}, causes relaxation (Figure 7.6a). By combining the flash photolysis procedure with a freeze-clamping device, the time course of ATP hydrolysis in a muscle fibre has been evaluated [73]. A burst of Pi formation is observed with which has very similar kinetics to that of isolated S1 in solution (cf. Figure 5.2). The final state is predominantly the M.D.P species, in agreement with earlier low-time resolution experiments [74]. Thus the properties of the detached heads in permeabilized muscle fibres appear reasonably similar to those deduced from solution studies.

In the presence of Ca^{2+}, photolysis of caged-ATP causes a fibre, initially in rigor, to lose tension transiently before regaining the full isometric value (Figure 7.6 b). The rate of the first phase is dependent on [ATP] and is attributed to the dissociation of the A.M complex [see Goldman, 1987]. The observed second order ATP binding rate constant is about 5×10^5 M^{-1} s^{-1}, only marginally slower than that in solution (section 5.3). At the physiological [ATP] of 4 mM, the effective rate constant for the dissociation of the A.M state by ATP would be about 2000 s^{-1} and therefore would not limit the ATPase cycle. Furthermore this rate constant is little affected by changing the initial tension of the muscle. The second phase observed after photolysis describes the formation of active tension-generating states and has a similar rate constant to the regain of tension described in the experiment of Figure 7.4, when performed under similar conditions. Pi affects the kinetics of the second phase and reduces its amplitude as expected from the steady-state measurements reported above [see Homsher and Millar, 1990]. These studies confirm that the tension-generating states involve product complexes, while the ATP binding and hydrolysis steps are little influenced by the myofibril lattice.

7.5 STRUCTURAL STATES OF CROSSBRIDGES

Having characterized chemical states in muscle fibres we now return to the question of their structural properties. Addition of ADP to rigor muscle has only a marginal effect on tension and the appearance of the crossbridges in the electron microscope. Other nucleotide analogues have therefore been sought in an attempt to trap the ATPase in different states in the cycle [see Cooke, 1986]. The non-hydrolysable analogue, β-γ imido ATP (AMPPNP), was developed in an attempt to trap an M.ATP-like state, however, the complex formed with myosin does not behave as a weak binding state in solution. Addition of AMPPNP to rigor insect flight muscle causes changes in the X-ray pattern and electron microscope image which suggest that it may partly reverse the crossbridge cycle. When a muscle held at fixed length goes into rigor it generates tension during the final ATP turnover. This tension is reduced to zero, if the muscle is allowed to shorten by about 1%, indicating that the final throw of the crossbridge is of the order of 13 nm (i.e. 1% of the half-sarcomere length). The rigor tension is also reduced on AMPPNP addition, although the stiffness remains high [75]. Stretching the muscle by 0.15%, which corresponds to a 2 nm displacement of a crossbridge, re-establishes the original value of the tension. This small length change was exploited in a mechanochemical coupling experiment in which an insect fibre was alternately stretched and released in two separate baths containing AMPPNP. Initially the [AMPPNP] in each bath was identical, but after repeated operations the one in which the muscle was released became more concentrated [76]. This experiment demonstrates the reciprocity of the mechanochemical energy transduction process, i.e. stretching a muscle favours a state which binds nucleotide more tightly (Figure 7.7).

The interpretation of the electron micrographs in the presence of AMPPNP proved difficult because the effects seen were very temperature dependent [77]. There was an increase in the appearance of projections at 90°, although these coexisted with the characteristic rigor chevrons. One possible interpretation is that one head of each myosin molecule detaches and takes up a relaxed position, while the other head remains attached and contributes to the high stiffness. However, addition of

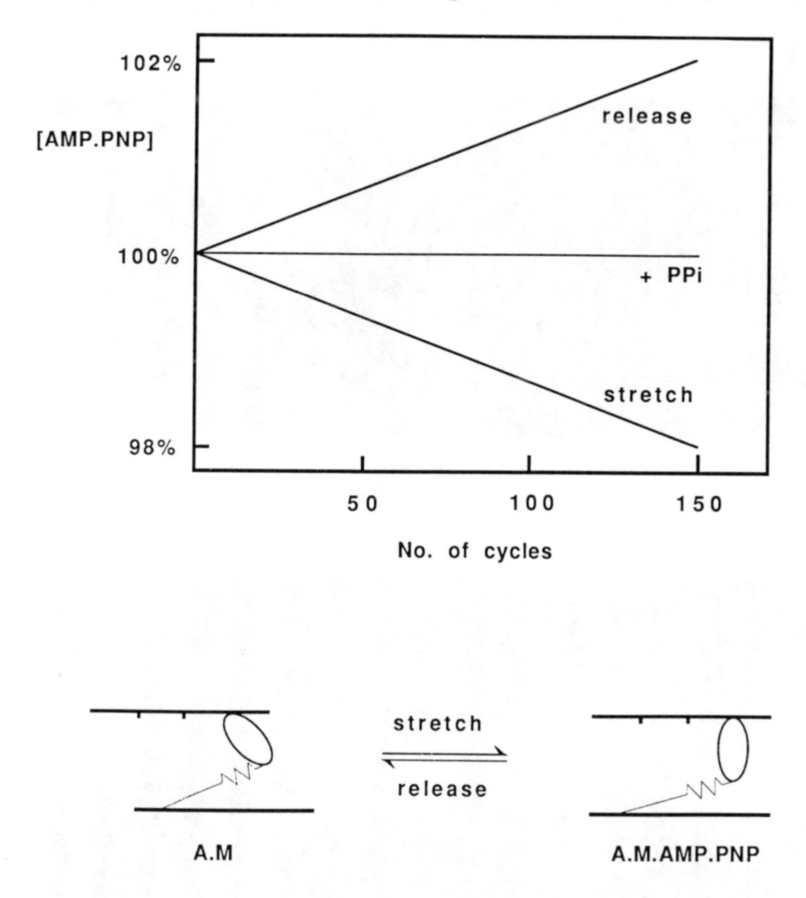

Fig. 7.7 Mechanochemical coupling in insect flight muscle. A skinned fibre was alternately stretched in a bath initially containing 96.5 μm AMP.PNP and then released in a second bath containing the same [AMP.PNP]. After many cycles the AMP.PNP became depleted in the first bath and concentrated in the second, owing to the preferential binding of AMP.PNP to the strained state. Inclusion of 2 mM pyrophosphate (PPi), a competitive inhibitor, abolished the effect, showing it was not caused by non-specific lattice forces. (After Kuhn [76].)

ethylene glycol had a synergistic effect with the AMPPNP and at a critical concentration, its addition seemed to convert the majority of the heads to a 90° state, although the stiffness remained high thus arguing that most were attached (Figure 7.8). AMPPNP plus ethylene glycol above this concentration causes relaxation, practically identical to that produced by

(a)

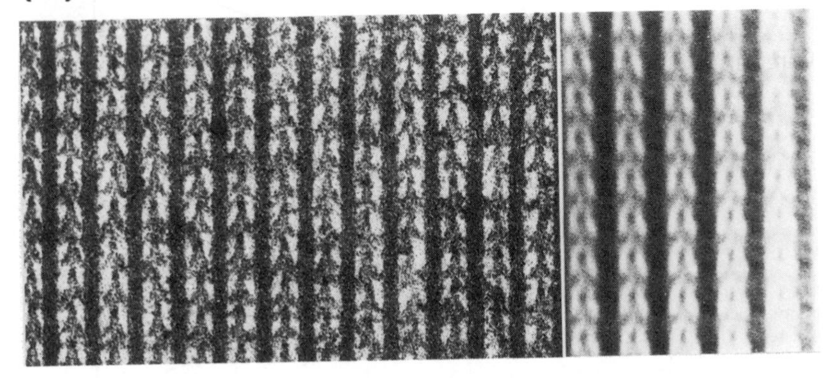

(b)

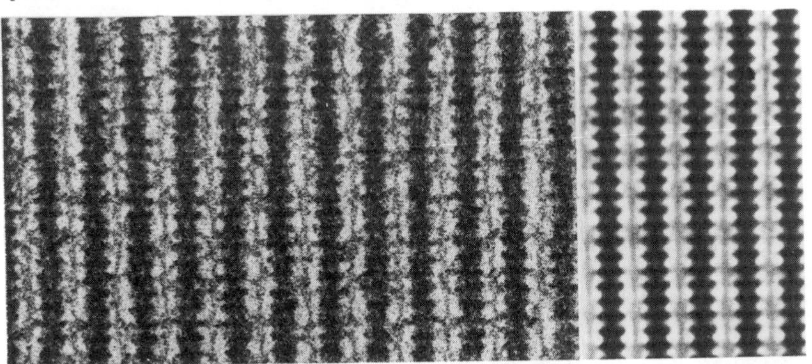

Fig. 7.8 Electron micrographs and corresponding filtered images of insect flight muscle (a) in rigor and (b) in the presence of 1 mM AMP.PNP and 30% ethylene glycol. Mechanical studies show the muscle remains stiff in (b) indicating that the crossbridges are attached although they take on angular positions similar to those of relaxed muscle. (Courtesy of M.C. and M. K. Reedy [77].)

ATP itself. These experiments provide the best evidence for the potential for some kind of angle change or a bending within the attached bridge, so accounting for the change in the dominant periodicity [77]. However, these observations must still be regarded as being indirect evidence for a change in crossbridge angle during contraction, particularly as the magnitude of the AMPPNP effect is only 10% that of the expected stroke size.

TOPICS FOR FURTHER READING

Cooke, R. (1986) The mechanism of muscle contraction. *CRC Critical Reviews in Biochemistry*, **21**, 53–118. (A comprehensive review.)

Goldman, Y.E. (1987) Kinetics of the actomyosin ATPase in muscle fibers. *Annual Review of Physiology*, **49**, 637–654. (Focuses on the use of caged compounds and isotope exchange.)

Hill, T.L. (1989) *Free Energy Transduction and Biochemical Cycle Kinetics.* Springer–Verlag, New York. (An introduction to the formalism of energy coupling.)

Homsher, E. and Millar. N.C. (1990) Caged compounds and striated muscle contraction. *Annual Review of Physiology*, **52**, 875–896. (Covers both contraction and activation mechanisms.)

Brenner, B. (1990) Muscle mechanics and biochemical kinetics, in *Molecular Mechanisms in Muscular Contraction*, (ed. J.M. Squire), Macmillan Press. (A thorough and lucid account of mechanical and chemical states in muscle.)

8 Molecular basis of regulation

8.1 THICK AND THIN FILAMENT REGULATION

Early studies on skeletal muscle actomyosin showed that the accessory proteins necessary for regulation of ATPase activity by Ca^{2+} were located on the thin filament (section 4.5). Subsequently, studies on other muscles, which apparently lacked troponin, revealed regulatory systems associated with the thick filament. A simple test was devised to determine the mode of regulation in any particular type of muscle based on the fact that, in the absence of additional regulatory proteins, rabbit skeletal muscle actomyosin shows maximal ATPase activity regardless of the Ca^{2+} concentration [78]. A crude myofilament preparation from the muscle in question is first prepared by blending, and the presence of a regulatory system is established by comparing its ATPase activity in the presence and absence of Ca^{2+}. The preparation is then mixed with either rabbit myosin or purified rabbit actin (i.e. lacking troponin and tropomyosin) and the Ca^{2+} sensitivity of the ATPase activity of the mixtures is determined. If the preparation is thick-filament regulated but lacks thin-filament regulation, its actin will interact with the rabbit myosin to give a high ATPase level independent of Ca^{2+}. If the preparation is thin-filament regulated but lacks thick-filament regulation, its myosin will interact with the rabbit actin to give a high, Ca^{2+}-independent ATPase activity.

By this criterion, molluscan striated muscles were found to be thick-filament regulated but lacked thin-filament regulation. On the other hand, most other invertebrates were dually regulated. Vertebrate skeletal muscles appeared to be solely thin-filament regulated. Although this test allowed a broad classification, controversies remained because some regulatory systems may be labile. The test was also ambiguous when phosphorylation mechanisms were involved, since the result depended on the level of contaminant kinases and phosphatases. Nevertheless, these studies did highlight the diversity of regulatory mechanisms in muscle, compared with the consistency of actin and myosin as the energy-transducing motor proteins. In this chapter, several types of regulatory mechanisms are discussed for which information is available at the molecular level.

8.2 REGULATION OF VERTEBRATE SKELETAL MUSCLE

As alluded to earlier, reconstitution experiments demonstrated that the Ca^{2+} sensitivity of purified skeletal muscle actomyosin could be achieved by adding back tropomyosin and troponin (section 4.5). Addition of tropomyosin alone can either activate or inhibit the actomyosin ATPase, depending on the conditions, but the system is not dependent on [Ca^{2+}]. Fractionation of troponin yielded three subunits: TnI, TnC and TnT [30]. Addition of TnI inhibited the actomyosin ATPase, regardless of the [Ca^{2+}]. Addition of TnC counteracted the inhibition of TnI, but now the ATPase was fully active again, regardless of Ca^{2+}. Addition of TnT to complete the system returned the full function, i.e. the ATPase was low in the absence of Ca^{2+} and high in its presence [see Ohtsuki *et al.*, 1986, Chapter 4].

A plausible structural mechanism for regulation was derived from the changes in the X-ray diffraction pattern of intact muscle on activation. Tropomyosin lies along the long pitch helix of the actin filament (Figure 4.14), but its effect on the intensity of the actin layer lines depends on its azimuthal position. On activation, the intensity of the second layer line (18.5 nm) increases, suggesting that the tropomyosin moves towards the centre of the groove. The principle is illustrated in Figure 8.1. This finding, combined with the estimated position of tropomyosin and myosin in low resolution electron micrographs of decorated thin filaments, led to the steric blocking model. In the absence of Ca^{2+}, troponin was

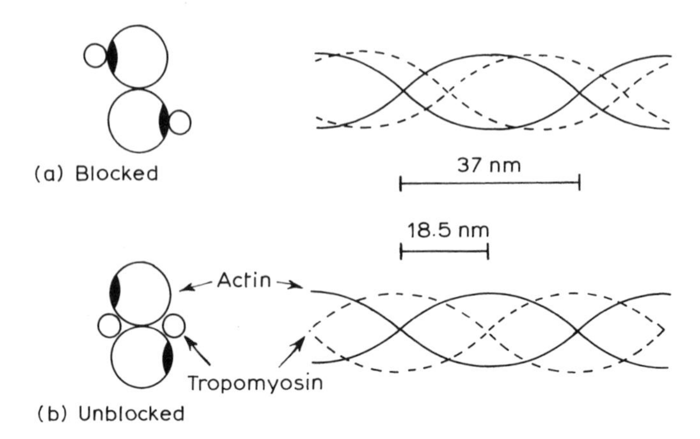

Fig. 8.1 The movement of tropomyosin on activation and its possible role in 'steric blocking'. In cross-section, azimuthal movement of the tropomyosin into the actin groove unblocks the myosin binding site. In side-on view this movement appears as a phase shift in the actin and tropomyosin long pitch helices and causes an intensification of the 18.5 nm periodicity.

proposed to hold tropomyosin in its blocking position towards the edge of the groove, so preventing myosin interaction. The binding of Ca^{2+} to TnC transmits a message, via TnI and TnT, to tropomyosin, inducing it to move into the groove and expose the myosin binding site on the actin filament [see Leavis and Gergely, 1984]. Time-resolved X-ray studies on frog muscle indicated that the putative tropomyosin movement occurred ahead of crossbridge movement, as indicated by the equatorial reflections, which in turn were ahead of tension development [79]. Thus the model appeared kinetically competent.

Biochemical studies led to a modification of this model. At low ionic strengths, at least, S1 binding to regulated actin filaments shows little dependence on Ca^{2+} [80]. Hence Ca^{2+} does not appear to affect step k_{-c} (equation 5.8), as would be expected from the steric-blocking mechanism. Subsequent studies with heavy meromyosin revealed a five- to ten-fold dependence of the binding constant on Ca^{2+}, but this is still an order of magnitude less than Ca^{2+}-dependence of the ATPase. It was suggested, therefore, that tropomyosin inhibits a transition in the myosin head from a weak to strong actin binding state (sections 5.3 and 7.3), rather than blocking binding to actin itself. Furthermore, full activation of the ATPase was only observed after a threshold of S1 heads had

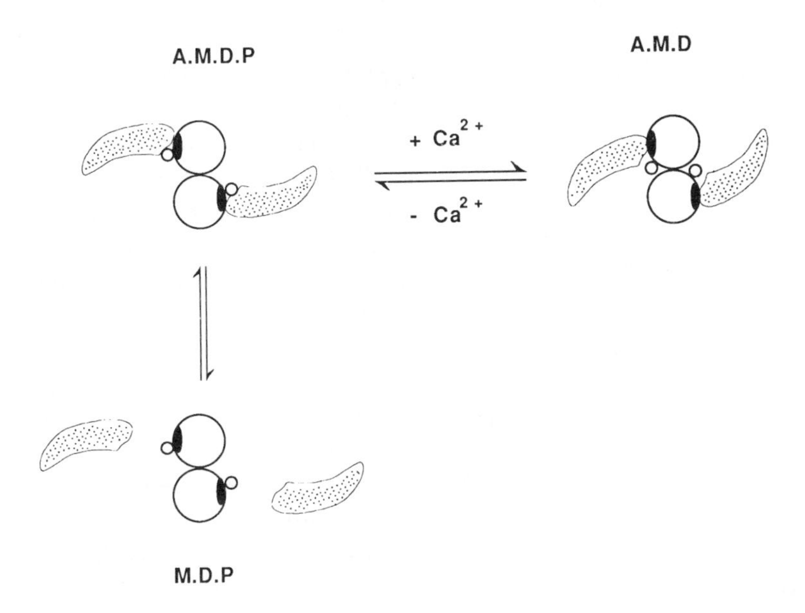

Fig. 8.2 A modified 'steric blocking' model in which tropomyosin blocks the transition from a weak- to a strong-binding state of the myosin head in the absence of Ca^{2+} and presence of ATP. In the absence of ATP, the strong-binding A.M state can push the tropomyosin into the 'relaxed' position.

attached to actin [81]. Thus tropomyosin movement into the actin groove appears to involve both a pull from troponin and a push from S1. Higher resolution analysis of electron micrographs of decorated thin filaments (Figure 4.13) indicates that the S1 and tropomyosin sites on the actin filament only partially overlap [27]. This is consistent with the idea that S1 can bind weakly to actin, via a limited attachment; then, provided Ca^{2+} is present, it can move into a strong binding position in concert with tropomyosin movement (Figure 8.2). In muscle fibres, a small degree of regulation of attachment of the weak binding states by Ca^{2+} could account for the X-ray data described above [see Brenner, 1990, Chapter 7].

TnC may be selectively removed from muscle fibres and replaced with fluorescently tagged molecules. Such an approach allows the kinetics of Ca^{2+} binding to be followed within the fibre. In addition, contraction in permeabilized cells may be rapidly initiated by the photolysis of Ca^{2+} chelators, such as nitr-5. With frog muscle, the time course of tension generation is similar to that of intact muscle stimulated electrically and indicates that Ca^{2+} release from the sarcoplasmic reticulum is not rate limiting *in vivo* [see Ashley *et al.*, 1991]. Other chelators, such as diazo-2, increase their affinity for Ca^{2+} on photolysis, and may be used to follow the kinetics of relaxation. Again the time course is comparable to that seen with intact muscle, showing that the Ca^{2+} pumping mechanism is not the limiting factor, at least in the case of frog skeletal muscle.

The activation of both ATPase and tension is highly cooperative so that only a small increase in Ca^{2+} is required for near maximal activation. This is, in part, due to the fact that each tropomyosin molecule controls several actin monomers. An interesting artefact is seen at low [ATP] *in vitro*, where some myosin heads may bind to form rigor complexes and keep the actin filament turned on in the absence of Ca^{2+} by displacing the tropomyosin. Thus the ATPase activity becomes Ca^{2+} independent at low [ATP]. This is also apparent in the contraction of permeabilized muscle fibres, in the absence of Ca^{2+}, because slow diffusion causes a reduced [ATP] in the centre of the preparation. At low [ATP], the first heads to dissociate may undergo a contraction cycle because other heads are still attached and render the tropomyosin inoperative [72].

There has been a long debate as to whether vertebrate skeletal muscle shows regulation at the thick filament level [see Rüegg, 1986]. One class of its light chains is phosphorylated by a specific kinase, which in turn is under the Ca^{2+} dependent control of calmodulin. The level of light chain phosphorylation has been shown to increase in intact muscle fibres during a tetanic contraction, although the time course of this process suggests that it plays a modulatory role rather than acting as a trigger for activating the muscle. Stimulating a muscle to twitch, following a brief tetanic contraction, gives a higher peak tension than a twitch produced by a fully relaxed muscle (so-called post-tetanic stimulation); a phenomenon whose time dependence is compatible with control by phosphorylation. Other studies have related phosphorylation to a decrease in ATP turnover during a maintained tetanic contraction [see Rüegg, 1986].

More controversial is whether the myosin can respond directly to Ca^{2+} binding, so activating crossbridges directly. The argument, in part, has been fuelled by the finding of myosin-linked regulation in molluscan striated muscles (section 8.3). However, when a vertebrate skeletal muscle is stretched to beyond overlap, stimulation does not produce any large changes in the equatorial and 42.9 nm layer line associated with the crossbridge position [82]. This result suggests that in a relaxed muscle most of the heads are close to the thick filament and are unaffected by Ca^{2+} directly. However, presumably each head occasionally and spontaneously explores the thin filament through thermal fluctuations and if an actin site is found, which is not blocked by tropomyosin, it may undergo a contraction cycle. This process then prevents the tropomyosin from blocking adjacent actin sites and so crossbridge binding escalates in a cooperative fashion leading to net contraction. In the absence of Ca^{2+} an exploring myosin head does not go through the complete contractile cycle and is likely to return to the thick filament backbone before other heads can find adjacent actin sites. The muscle therefore remains in a relaxed state.

8.3 REGULATION OF MOLLUSCAN MUSCLE

Although the regulatory mechanism found in molluscan muscle may prove to be a rather specialized example, its implications have had an important bearing on the evaluation of regulatory mechanisms in both striated and smooth muscle. Myofibrils prepared from molluscan muscles contain very little troponin, at least after washing at low ionic strength, yet their ATPase activity is Ca^{2+} dependent. Moreover, myosin from these muscles exhibits a Ca^{2+}-sensitive ATPase when activated by purified actin. The mechanism by which this regulation is achieved was revealed using scallop myosin which, on washing with EDTA, loses one type of light chain (the so-called regulatory light chain) and becomes insensitive to Ca^{2+}[83]. The Ca^{2+}-sensitivity of the ATPase is restored when this light chain is added back to the desensitized myosin in the presence of Mg^{2+}. In contrast to vertebrate striated myosin, molluscan myosin binds two Ca^{2+} specifically and these sites are essential for the regulatory activity. Neither the isolated regulatory light chain nor desensitized myosin bind Ca^{2+} with high affinity and specificity. The regulatory light chain (RLC) does have a non-specific Ca^{2+}/Mg^{2+} site but this is common to all light chains of this class, including the rabbit skeletal phosphorylatable light chain. The kinetics of Ca^{2+} binding to the non-specific site are too slow for it to be involved directly in the activation process. It is possible that the Ca^{2+}-specific site is located on the other class of light chain (termed essential, for historical reasons), but the presence of the regulatory light chain is necessary for the proper folding of this subunit to give a functional site. A small proteolytic fragment containing the two light chains and the neck region of the heavy chain binds Ca^{2+} specifically and shows the same quench in tryptophan fluorescence as the intact myosin. The latter reflects a conformational change which is ultimately transmitted to nucleotide and actin binding sites [84].

The actin-activated ATPase activity of desensitized myosin preparations is permanently high. Thus the regulatory light chains are inhibitory subunits in the native myosin and activation by Ca^{2+} can be viewed as counteracting this inhibition. The regulatory light chains may also be selectively removed from scallop glycerinated muscle fibres by EDTA treatment, and this leads to a preparation that develops close to isometric tension, on addition of ATP, regardless of the Ca^{2+} concentration.

The light chains (RLC and ELC) are located near the S1–S2 junction of each myosin head. S1 can be prepared containing both light chains, but its actin-activated ATPase is Ca^{2+}-insensitive, unlike HMM which behaves as the parent myosin. The importance of the neck region in regulation has also been deduced from electron microscope studies.

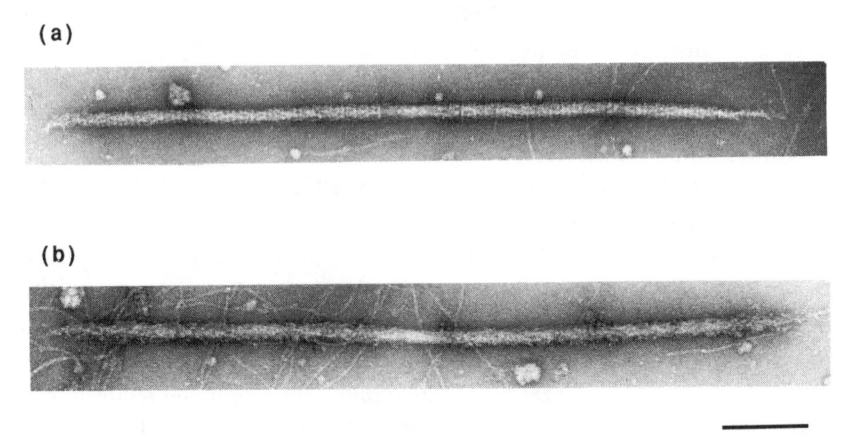

Fig. 8.3 Electron micrographs showing the effect of Ca^{2+} on the crossbridge disposition in scallop striated adductor muscle thick filaments. (a) Under relaxing conditions (absence of Ca^{2+}, presence of ATP) cross-striations with a 14.5 nm periodicity are apparent from the ordered myosin heads lying close to the filament backbone. (b) On addition of Ca^{2+} the heads splay out to give a fuzzy outline with no detectable order. (Courtesy of P. Vibert and R. Craig [85].)

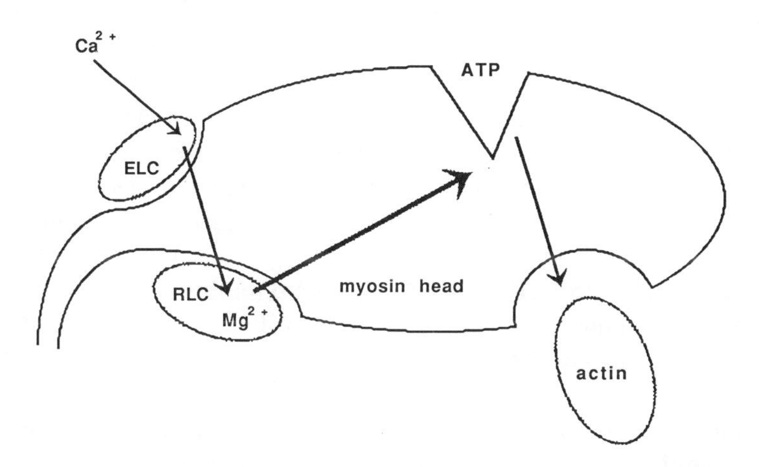

Fig. 8.4 A possible pathway of communication by which Ca^{2+} binding to the neck region of scallop myosin controls crossbridge interaction with actin. The location of the Ca^{2+}-specific binding site is tentatively identified as the essential light chain, although the regulatory light chain is vital for a functional site.

Native and synthetic thick filaments from scallop muscle show a regular 14.5 nm spacing of the crossbridges, as with other invertebrate thick filaments (Figure 8.3), provided ATP is present and Ca^{2+} absent (i.e. under relaxing conditions). Addition of Ca^{2+} to samples on the electron microscope grid causes a rapid loss of this order as the crossbridges splay out [85]. Removal of the regulatory light chain gives a similar appearance. Thus the activation mechanism in the intact muscle could involve the direct mobilization of the crossbridges from the shaft of the thick filament (cf. vertebrate skeletal muscle where crossbridge movement was not detected in filaments stretched beyond overlap; section 8.2). However, biochemical studies show that this is not the only consequence of Ca^{2+} binding to scallop myosin.

In typical steady-state assays, the actin-activated ATPase of scallop myosin is elevated about 20-fold by Ca^{2+}. However, even in the absence of actin, Ca^{2+} activates the ATPase by some five-fold. The significance of this apparently limited activation of myosin alone became clear when single turnover ATPase measurements were made on HMM and revealed two populations (see section 5.1).

The ATPase of 80% of the HMM was extremely suppressed in the absence of Ca^{2+}, so that the steady-state ATPase was dominated by the remaining 20% Ca^{2+}-insensitive population. The regulated population could be activated more than 600 fold by actin in the presence of Ca^{2+}, which is close to the degree of control that would be expected for the intact muscle. Subsequently it was shown that Ca^{2+} regulates the product release steps of the ATPase of both the myosin alone and its actin-activated state [86]. Taken in conjunction with limited proteolysis results [84], the path of communication from the Ca^{2+} binding site to the actin site was deduced as shown in Figure 8.4. It is not clear to what extent the neck region (or the second head) is required for this process. Thus Ca^{2+} controls actin interaction via events at the ATPase site, as also deduced for vertebrate striated muscle (section 8.2).

The ability to dissociate reversibly the regulatory light chain from scallop myosin enables light chains from other species to be tested for their ability to resensitize the ATPase to Ca^{2+}. Initial studies showed all myosins contain an equivalent RLC capable of binding to scallop myosin, but only those from smooth muscle and most invertebrate muscles were competent at fully resensitizing the ATPase and restoring Ca^{2+} specific binding. Hybrids containing two vertebrate skeletal RLCs were Ca^{2+}-insensitive and had a permanently inhibited ATPase activity. Overall, the results of these studies suggest that the regulatory light chains from vertebrate striated muscle are not competent to function as regulators through the direct Ca^{2+} binding mechanism but they retain a potential inhibitory function.

Molluscan muscle regulation is also of interest with respect to the catch mechanism. Phosphorylation of both the RLC and heavy chain of scallop smooth muscle myosin has been detected, and might be directly implicated in this process.

8.4 REGULATION OF VERTEBRATE SMOOTH MUSCLE

Early studies on smooth muscle suggested the regulation was myosin-linked and was achieved through the phosphorylation of the regulatory light chain [see Rüegg, 1986; Sellers and Adelstein, 1987]. Unlike striated muscle, phosphorylation has a large activating effect on the actin-activated ATPase *in vitro*. The myosin light chain kinase is activated by Ca^{2+} in the presence of calmodulin. In this way events in the smooth muscle cell which lead to an increase in cytosolic [Ca^{2+}] may be coupled to activation (e.g. binding of hormones to the α-receptors). However, smooth muscles can also be controlled by external effectors through the β-receptors which leads to an increase in cyclic adenosine 3′,5′-monophosphate (cAMP). This has the effect of rendering the myosin light chain kinase less active, through phosphorylation by cAMP-dependent protein kinase. As a result, in the whole organism, different smooth muscles may react differently to a common stimulus according to the receptors on their cell surfaces. For example, in the flight or fight response, adrenalin causes blood vessels to the gut to contract, to give that well-known sinking feeling, while the coronary arteries dilate to achieve a better oxygenation of the heart muscles and prepare the body for action.

The effects of RLC phosphorylation of smooth muscle myosin are two fold. It enables actin to activate the myosin ATPase. This is achieved through an enhancement of the product-release steps, rather than by changing the actin affinity (as with striated muscles described above). Phosphorylation also makes monomeric myosin less soluble [87]. For many years it was difficult to detect myosin filaments in relaxed smooth muscle [11]. This was due to their solubility under the relaxing conditions used during their preparation for microscopy (i.e. when they are dephosphorylated). Thick filaments have been observed, however, by careful preparation (Figure 3.7). Nevertheless, it is possible that in some smooth muscles, solubilization of the thick filament occurs *in vivo* during prolonged relaxation. The mechanism of this effect has been well studied *in vitro* where it has been found that dephosphorylated smooth muscle myosin has a tendency to fold up in a compact monomeric form. The tail bends at two points and contacts the neck region to give a molecule with a sedimentation coefficient of 10S, instead of 6S found for the normal extended form. The folded form traps ATP to give a long-lived M.D.P state which is inert with respect to actin activation and filament assembly [88]. This molecular species has also been characterized for non-muscle myosins, where assembly-disassembly appears to be important in different phases of the cell cycle (Figure 8.5). Thus contraction may be regulated by the control of the assembly of thick filaments in local regions of the cell. Myosin heavy chain phosphorylation at the tip of the tail has also been detected in smooth and non-muscle cells, and likewise appears to play a role in the control of filament assembly [89].

Fig. 8.5 The folded (10S) conformation of thymus myosin observed under relaxing conditions. As with smooth muscle myosin, this state is favoured when the regulatory light chains are dephosphorylated. (Courtesy of Dr J. Kendrick-Jones.)

Smooth muscle is also regulated at the level of the thin filament through an inhibitory protein termed caldesmon. The stoichiometry of caldesmon is low, but its effect may be amplified through tropomyosin. The Ca^{2+}-binding protein which presumably interacts with caldesmon to relieve the inhibition on activation has not been fully characterized, although calmodulin will achieve a partial restoration *in vitro*. More recently, another thin-filament based protein, calponin, has been implicated in the mechanism of regulation of smooth muscle contraction [90].

TOPICS FOR FURTHER READING

Leavis, P.C. and Gergely, J. (1984) Thin filament proteins and thin filament-linked regulation of vertebrate muscle contraction. *CRC Critical Reviews in Biochemistry*, **16**, 235–305. (Comprehensive, but written before regulatory proteins had been crystallized.)

Rüegg, J.C. (1986) *Calcium in Muscle Activation: a Comparative Approach*, Springer–Verlag, Berlin, Heidelberg, New York. (A detailed monograph on regulatory mechanisms.)

Sellers, J.R. and Adelstein, R.S. (1987) Regulation of contractile activity, in *The Enzymes* XVIII, (eds. P.D. Boyer and E.G. Krebs), Academic Press, Orlando and London, pp. 381–418. (Reviews phosphorylation mechanisms in smooth and non-muscle cells.)

Ashley, C.C., Mulligan, I.P. and Lea, T.J. (1991) Ca^{2+} and activation mechanisms in skeletal muscle. *Quarterly Review of Biophysics*, **24**, 1–73. (Reviews approaches for studying the kinetics of activation in striated muscle fibres.)

9 *In vitro* motility assays

9.1 TECHNICAL DEVELOPMENTS

The preceding chapters show that much of our understanding of the molecular mechanism of muscle contraction has come from trying to marry protein chemical information obtained in solution with physiological information obtained from fibres. Solution studies have the benefit of dealing with known components, but neither force nor movement is measurable. It has been a long-standing goal of muscle research to reconstitute the contractile system from the individual components, but in this regard little progress has been made beyond the actomyosin thread, developed in the 1940s (section 3.5). In the last decade, however, advances in optical technologies, coupled with other ingenuities, have allowed force and movement to be quantitated at the level of isolated proteins and filaments. Thus biochemistry and physiology are beginning to meet on common ground with these *in vitro* motility assays.

Actomyosin threads develop about 25% the tension of muscle fibres, when normalized for the protein content, but they contract with a very low velocity which shows no correlation with the myosin type. Nevertheless, such preparations have been used to show that single-headed myosin supports about half the tension of intact double-headed myosin, and hence these experiments argue for the functional independence of the heads [91]. However, it is difficult to quantitate contractile activity of actomyosin threads at the level of individual filaments because the majority of them are disordered. New assays are therefore required.

A good correlation between contractile velocity *in vivo* and *in vitro* is observed in an *in vitro* assay which exploits the natural actin filament bundles found in giant algal cells, such as in *Nitella* and *Chira*, and synthetic beads coated with myosin [see Scheetz *et al.*, 1986]. In such cells, cytoplasmic streaming is readily observed under the light microscope, as evident by the movement of organelles around the cell perimeter at velocities of up to 70 μm s^{-1}. The actin cables can be accessed by surgically splitting the cell. Polystyrene beads, which have been coated with myosin, adhere to the actin cables in the absence of ATP. In the presence of ATP, they move with a steady velocity dependent on the source of the myosin (about 5 μm s^{-1} for rabbit skeletal myosin at 25°C). The assay is limited in that the properties of the thin filaments cannot be explored in this system. It is possible to obtain bead movement on purified muscle actin, attached to a substrate and oriented by flow, but quantitation proved difficult and other assays have superseded this approach. These require more sophisticated applications of light microscopy.

Although the light microscope can only resolve objects which are separated by at least $\lambda/2$ (i.e. > 200 nm), it can detect particles much smaller than this. Single actin filaments which have been labelled with a fluorophore can be observed by eye through the microscope, but their apparent width is about 200 nm as determined by the above diffraction limit. Using a video camera as a detector allows the contrast of the image to be greatly improved (Figure 9.1). Contrast enhancement is possible because the video detector has a much wider dynamic range than the human eye or photographic film. Thus the image may be recorded using high light intensities and the background level subtracted electronically. Contrast enhancement also allows non-fluorescent objects, smaller than the diffraction limit, to be detected from their scattering characteristics. Myosin filaments, microtubules and synthetic microbeads can all be visualized using video-enhanced differential interference contrast (DIC) microscopy [see Amos and Amos, 1991, Chapter 1]. Another method for imaging small organelles with good contrast is confocal reflection microscopy. In favourable cases, displacements of the object can be calculated with a precision of about a nanometre by calculating the centroid position of the image. Standard video allows frames to be captured at 25 to 30 s^{-1}. These specifications are approaching those required to visualize an individual crossbridge in action [see Spudich, 1990].

Fluorescently-labelled actin filaments are observed in the microscope

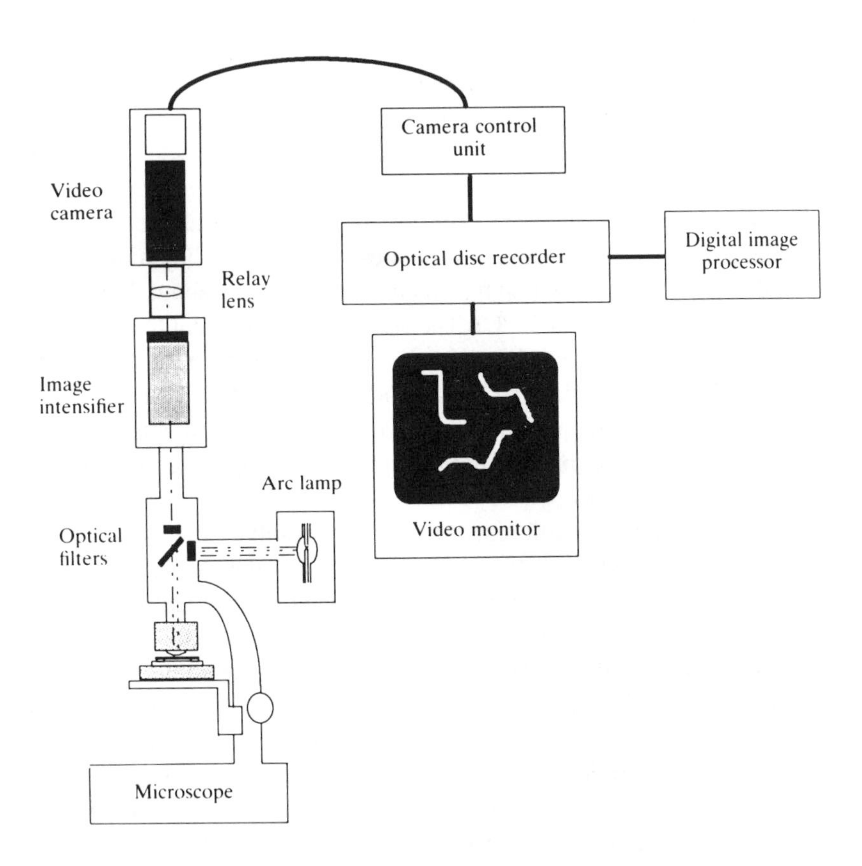

Fig. 9.1 Schematic diagram of a video-enhanced, epifluorescence microscope used to detect the movement of phalloidin-labelled actin over a myosin-coated cover slip. The dichroic mirror in the microscope reflects light at the excitation wavelength (540 nm) but transmits the emitted light (580 nm) to the detector. [From Kron *et al.* (1991).]

to undergo continuous thermally-driven flexing movements. In the presence of ATP and S1, this motion increases in frequency and amplitude. More dramatic, however, is the movement observed when the myosin molecules are immobilized on a glass surface (Figure 9.2). Now actin filaments in the plane of the immobilized myosin are observed to slide their way across the field at velocities of about 5 $\mu m\ s^{-1}$ (for rabbit skeletal myosin [92]). The actin is usually visualized by labelling with phalloidin which has been derivatized with the fluorophore, rhodamine [see Kron *et al.*, 1991]. Phalloidin binds tightly to each subunit in F-actin and

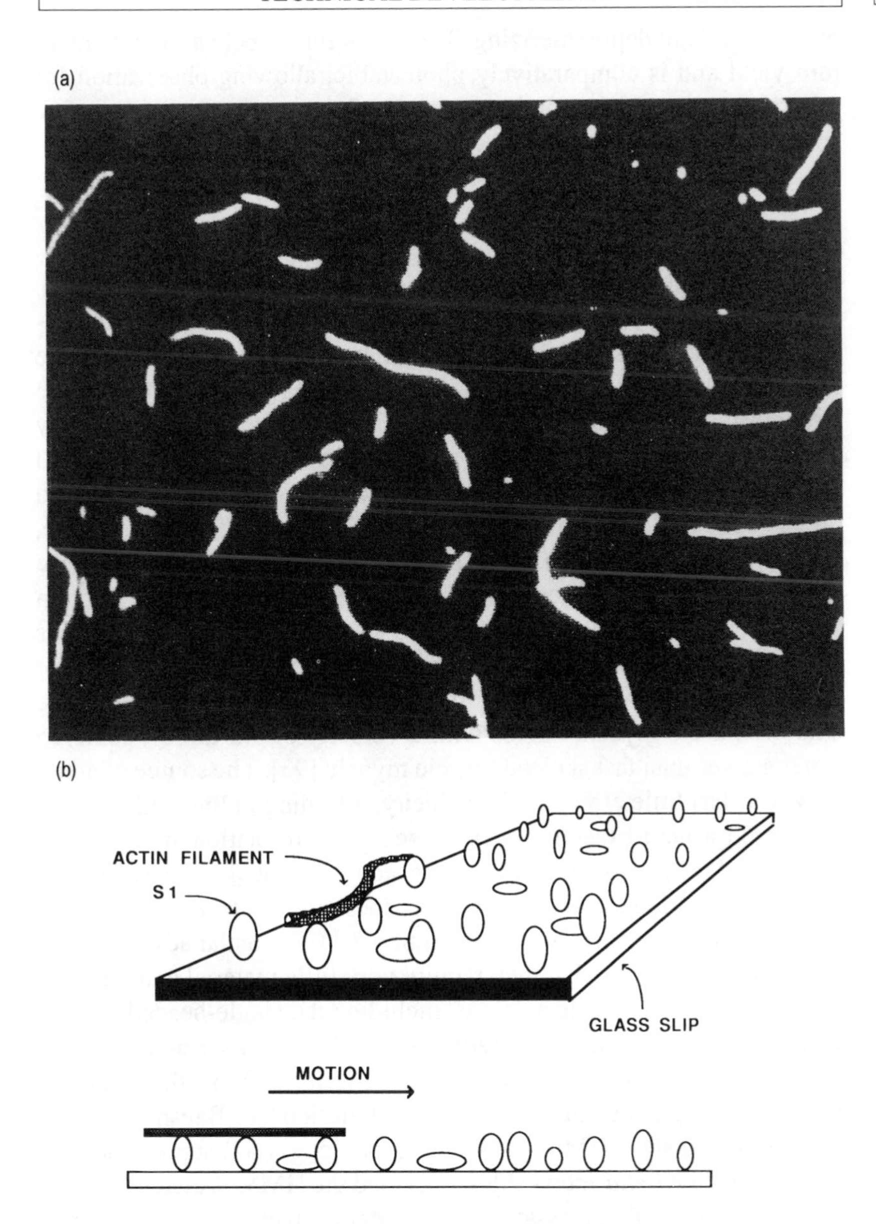

Fig. 9.2 (a) Image of phalloidin-labelled actin filaments observed by epifluorescence microscopy [From Kron *et al.*, (1991).] (b) Schematic diagram of actin filament movement on a myosin S1 surface. (Courtesy of Dr R. Ankrett.)

prevents it from depolymerizing. The rhodamine label has a high quantum yield and is comparatively photostable, allowing observation for several minutes. In favourable cases the fluorescence from just a few rhodamine molecules has been detected.

9.2 FACTORS REQUIRED FOR SLIDING

In vitro motility assays show a strong dependence on ionic strength. Above 0.1 M salt, ATP addition causes myosin-labelled beads to dissociate from *Nitella* actin and fluorescent actin filaments to dissociate from a myosin surface, so no further interaction is possible. At low ionic strength (20–50 mM) the velocity of sliding shows an approximately hyperbolic dependence on [ATP] with an apparent K_m which is about 10 times higher than that for the actin-activated ATPase activity, measured under comparable conditions. It is probable that the long-lived A.M state, which would predominate at low [ATP], causes considerable resistance to sliding, as noted for the contraction of skinned muscle fibres (section 7.3). The observed velocity of sliding *in vitro* at saturating [ATP] shows a reasonable correlation with the maximum velocity of shortening, V_o, of the muscle type from which the myosin is extracted. Thus vertebrate smooth muscle myosin moves with a velocity of 0.2 μm s^{-1}, some 25 times slower than fast skeletal muscle myosin [93]. The source of actin, however, has little effect on the velocity, a finding in line with its very conserved structure (section 4.2). There is less correlation of the *in vitro* sliding velocity with the V_{max} of the actin-activated ATPase but, as discussed previously, this parameter does not relate directly to the velocity of the crossbridge stroke (section 7.1). Particular advantages of the *in vitro* assays are that they require very little material and they can be applied to non-muscle myosins, including the single-headed myosin I class, whose mechanical properties are difficult to determine *in vivo*.

A good deal of effort has been directed towards defining the minimum proteolytic fragment which would support motion [see Bagshaw, 1987]. Both long and short HMM (lacking the full length S2) support particle movement on *Nitella* actin cables, provided the HMM is attached via an antibody spacer. These assays do not function with S1. However, it is possible to observe actin filament motion over S1, when the latter is immobilized on a nitrocellulose surface [94]. S1 with an intact neck region, produced by papain digestion, is much more efficient than chymotryptic S1 in producing movement, even though their actin-activated ATPases in solution are comparable. Even so, the movement is about four times slower than observed with HMM. This may reflect the importance of the neck region in the mechanochemical coupling, or simply that the heads are in a more favourable orientation when attached to the substrate via the tail region. In either event, it seems clear that actin sliding does not require S2, and therefore the proposal of a helix–coil transition in this region as being the prime mover seems untenable (cf. section 6.5).

The direction of movement is controlled by the polarity of the actin filament. Myosin heads always move from the 'pointed end' towards the 'barbed end' as expected from their action in muscle (see section 4.4). In the *Nitella* assay, it was considered that myosin heads in a favourable orientation caused the bead to move, while those which happened to be in the wrong orientation failed to interact with the polarized actin cable beneath it. Somewhat surprising was the subsequent discovery that myosin molecules in the 'wrong' orientation can generate movement, presumably by twisting around the S1–S2 region, so that the precise stereochemistry of the S1–actin interface is maintained. This was clearly demonstrated by the observation that actin filaments can slide over immobilized thick filaments, cross the bare zone and continue to move on the other side [95]. The precise tracks taken by the actin can be correlated with the position of the thick filaments by switching between fluorescence and DIC detection during the assay.

Regulatory proteins are functional in the *in vitro* motility assays. When the actin filaments are reconstituted with troponin and tropomyosin, their movement over immobilized myosin requires the presence of Ca^{2+}. In the absence of Ca^{2+}, the actin tends to diffuse away from the surface. Thus the weak binding states are not sufficient to maintain a long-term interaction with the actin. Myosin-linked regulation can be demonstrated with both the *Nitella* assay and the actin filament assay. In the case of molluscan and vertebrate smooth muscle myosins, movement requires the presence of Ca^{2+} or light chain phosphorylation, respectively. Non-phosphorylated smooth muscle myosin does not move actin filaments, but at low ionic strength the resultant weak binding state is sufficient to keep the filaments from diffusing away from the surface [93].

9.3 FORCE MEASUREMENTS

The calculated resistive force against bead or filament movement due to the viscosity of the medium is more than an order of magnitude lower than that estimated to be produced by a single crossbridge (section 7.1). Thus in the assays described above, the filaments are sliding in a practically unloaded condition. It is therefore not possible from the above observations on filament sliding to rule out the active participation of other domains of the myosin molecule, such as S2, in the development of the full isometric force. How can force be measured in these *in vitro* motility assays?

It is not practical to increase the viscosity of the medium to a sufficient extent to impede motion. Filament motion is impeded by the presence of rigor heads (i.e. the A.M state) at low [ATP], but this effect cannot be accurately quantitated. A similar resistance is seen when a small fraction of the myosin population is modified with N-ethylmaleimide (NEM). This reagent labels the SH1 cysteine and produces a myosin head which does not dissociate from actin in the presence of ATP. Nevertheless, under these conditions, long filaments of actin tend to break into shorter fragments, suggesting that the active heads generate a significant force. Interestingly, crossbridges in the weak binding state are also capable of producing a resistive force. This has been demonstrated by investigating copolymers of phosphorylated and non-phosphorylated smooth muscle myosin [93]. When the phosphorylated myosin (i.e. the activated form) was less than 25% of a mixed polymer, filament sliding was retarded.

A quantitative estimate of the force produced by a small number (about 100) myosin heads interacting with a single actin filament was deduced from the bending of a fine glass microneedle (Figure 9.3). First, the needle was coated with NEM-labelled myosin, to form a specific ATP-resistant glue for the actin filament. Then the opposite end of the actin filament was manipulated to lie over a myosin- or S1-coated coverslip. On addition of ATP, the needle was observed to bend towards the cover slip in about 50% of the trials (it was not possible to deduce the actin filament polarity prior to the experiment [96]). By calibrating the compliance of the needle and estimating the number of interacting heads, it was deduced that each head could generate a force of about 0.5 pN, a value only three-fold lower than that calculated for intact muscle fibres (section 7.1). This is a very important finding because it indicates that

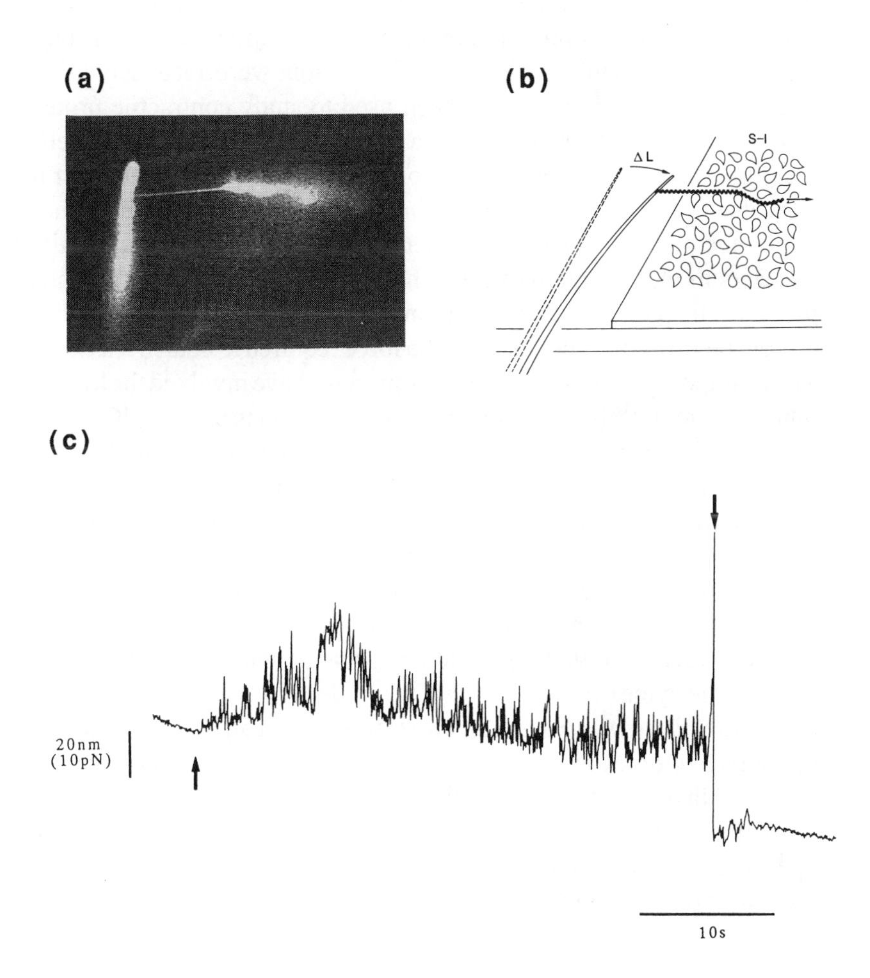

Fig. 9.3 (a) Image of a single phalloidin-labelled actin filament attached to glass microneedles. (b) Schematic diagram of the bending of a microneedle when an actin filament is lowered over a myosin S1 surface in the presence of ATP. (From Kishino and Yanagida [96].) (c) Fluctuations in the displacement of a stiff needle when an actin filament is lowered over a small number of myosin heads (↑). The actin filament detaches at the point marked ↓. (From Ishijima *et al.* [97].)

the S1 moiety, in combination with actin, is sufficient to generate considerable tension as well as rapid unloaded sliding. The crossbridge cycle predicts that as the number of heads interacting with the actin becomes small, the force should show random fluctuations because of the stochastic nature of the actomyosin cycle (Figure 9.3). This has been confirmed in recent studies under isometric conditions, although at high velocities of shortening the expected fluctuations were not observed [97].

Recently, optical traps have been used to study contractile proteins [see Spudich, 1990]. A particle in a laser beam tends to centre itself at the point of highest intensity due to forces from the momentum of the light itself. The force imposed is in proportion to the intensity and is within the pN range. This phenomenon can be used as a minute pair of tweezers to manipulate particles which have been coated with a motor protein. The particle can be positioned on the complementary filament protein (actin or tubulin) and then the force required to detach it assessed. The first applications with contractile proteins have involved the kinesin–tubulin system [98] and similar studies are now in progress with myosin and actin. Kinesin has some broadly analogous properties with myosin in regard to its ATPase activity, although during its sliding interaction with tubulin, the attached states seem to predominate (i.e. it has a higher duty cycle, cf. section 7.1). This may be related to its role *in vivo* where it transports small organelles capable of interacting with only a few kinesin molecules. In contrast, myosin heads in a muscle act in large numbers *in vivo*, so that only a few need be bound at any instant in the crossbridge cycle.

Overall, these *in vitro* motility assays represent an important advance in the technologies to study contractile proteins. The assays have demonstrated that the interaction of the S1 moiety with actin is sufficient for mechanochemical energy transduction. Nevertheless, they have also opened up some controversies which bring the classical crossbridge cycle into question (section 10.1).

TOPICS FOR FURTHER READING

Sheetz, M.P., Block, S.M. and Spudich, J.A. (1986) Myosin movement *in vitro*: a quantitative assay using oriented actin cables from *Nitella*. *Methods in Enzymology* **134**, 531–544. (Covers practical aspects of this *in vitro* assay.)

Bagshaw, C.R. (1987) Muscle contraction: are two heads better than one? *Nature*, **326**, 746–747. (A brief commentary.)

Spudich, J.A. (1990) Optical trapping: motor molecules in motion. *Nature*, **348**, 284–285. (Current and future applications of optical tweezers.)

Kron, S.J., Toyoshima, Y.Y., Uyeda, T.Q.P. and Spudich, J.A. (1991) Assays for actin sliding movement over myosin-coated surfaces. *Methods in Enzymology*, **196**, 399–416. (Practical aspects for following actin movement.)

10 Problems and prospects

10.1 THE CROSSBRIDGE CYCLE

There is no doubt that crossbridges are central to the molecular basis of muscle contraction because they are the location of the enzymic activity responsible for ATP hydrolysis. Nevertheless, the nature of their mechanical coupling with the actin filament during active contraction is unclear. The textbook diagram of a rigid oar-like crossbridge, which rows through a 90° to 45° angle, is not strongly supported by experimental evidence. A crossbridge which bends in the neck region accommodates some experimental findings, although other observations argue against any major change in conformation of the head. However, before attempting to pin down the details of the crossbridge cycle in terms of the shape or change in angle of attached S1, there is a broader question which needs to be resolved concerning the crossbridge stroke size, i.e. the distance the crossbridge moves when attached to actin for each round of ATP hydrolysis.

The worked example of the crossbridge stroke size given in section 7.1 indicated that a 10 nm displacement is thermodynamically and kinetically feasible, although the calculated fraction of attached heads at the maximum shortening velocity is rather small. This is of little consequence when dealing with the large numbers of heads working in concert in each half-sarcomere of the myofibril. To reduce the number of possible interactions, Yanagida and colleagues digested the Z-lines of crab myofibrils, leaving each thin filament to slide independently. They then measured the shortening speed by direct observation under a microscope and the ATPase activity of the preparation. From these data a stroke length of > 60 nm was obtained [99]. This value is clearly inconsistent with a cycle of the type illustrated in Figure 7.1 for a 20 nm crossbridge. Conditions were therefore sought using *in vitro* motility assays, where even fewer myosin heads could interact with any one actin filament, to check this result. The experiment has been undertaken in two ways and led to opposite conclusions.

Yanagida and colleagues analysed the velocity of actin filaments over a high density of immobilized myosin heads (average separation = 11 nm) and found that short actin filaments (40 nm) moved at the same speed as long filaments. Since a short filament must make contact with at least one head to sustain the observed V_o (the inertia is negligible) and to prevent the filament from diffusing away, this observation places a lower limit on the number of heads attached per unit length of actin. A 40 nm filament contains about 16 actin monomers, of which about half might be in a position capable of interacting with myosin heads immobilized on a surface. From the ATPase activity of the preparation measured under the same conditions, the stroke size, a, was calculated to be > 100 nm [100].

On the other hand, Spudich and colleagues reduced the average number of interaction sites by lowering the density of myosin heads on the substrate (head separation > 50 nm). Under the usual assay conditions, actin filaments below a minimum length tended to diffuse away from the surface presumably because the probability of an advancing actin filament finding a myosin head was low. Determination of this minimum length, in combination with the ATPase rate per unit filament length, suggested a stroke size in the range 8–40 nm. They went on to characterize the movement of shorter actin filaments further by inclusion of a polymer, methylcellulose, which markedly reduced the lateral diffusion of actin away from the surface, but caused little hindrance to its longitudinal movement. Short actin filaments now slid on low head density substrates, but the velocity became progressively slower as the head density was reduced. The low velocities were attributed to filaments which were driven by just a few heads (i.e. the actin filament was dissociated for most of the time but was kept in the vicinity of the heads by the methylcellulose and received occasional impulses). Calculations from this experiment led to a crossbridge stroke, a, of between 10 and 28 nm [101].

Apart from some differences in the raw data obtained, these experiments differ in their underlying assumptions. At high head densities, it is possible that an unloaded actin filament is driven predominantly by crossbridges at its leading end. In the presence of excess ATP, these heads will already be in the M.D.P state and could undergo an immediate crossbridge stroke and then dissociate to complete the ATPase cycle on a somewhat longer timescale (cf. section 5.3, where it was suggested that actin could interact with more than one myosin head during each complete cycle of the ATPase). The measured ATPase, however, reflects that of the whole population of actin filaments and might be dominated by the longer actin filaments where this 'end effect' will become negligible. On the other hand, the alternative approach using a low head density requires assumptions about the number of available active heads, in order to estimate the fraction which were attached. Further work is required to sort out these ambiguities [see Huxley, 1990].

The conventional crossbridge cycle of equation 7.1 is too rigid in its formulation, in that attachment and detachment are now known not to be obligatorily coupled to ATP hydrolysis (cf. equation 5.8). Furthermore, when the x-dependence of such schemes is considered (Figure 7.2b), it is conceivable that strongly-bound crossbridges (e.g. the A.M and A.M.D state) may be torn off during an unloaded shortening, under the influence of neighbouring crossbridges, and rapidly reattach to the actin filament several subunits along [66]. In this way the energy of ATP hydrolysis may be fractionated over several crossbridge cycles. This kind of behaviour may account for the decrease in ATP consumption at high velocities of shortening (section 2.5, [102]). However, the total crossbridge stroke for a model of the type shown in Figure 7.2a would not exceed the dimensions of the myosin head. An alternate mode of fractionation involves transfer of energy to the actin filament which, in turn, may drive the sliding of more than one head for each ATP hydrolysed. The considerable flexibility of the actin structure, which could serve as a temporary energy store, is suggestive of this idea.

The influence of weakly attached crossbridges also needs to be considered. Recently, interest has been rekindled in diffusion processes of biopolymers, following the discovery of the anomalously high rates at which some DNA binding proteins can locate a small region of specific base sequence. It is proposed that the protein may initially collide and form a weak attachment with any part of the DNA molecule, then undergo one-dimensional diffusion until the specific site is reached [103]. Such a process is extremely rapid and net displacements of hundreds of nanometres can be achieved on the millisecond timescale. The bonds involved in such attachment might be group specific but form very short-lived interactions, or long-distant electrostatic forces which function more as a continuum.

Experimental evidence that contractile proteins can undergo such thermally driven random displacements is provided by the analysis of the dynein ATPase (the motor protein of eukaryotic flagella). Microtubules are observed to slide rapidly backwards and forwards on a dynein-covered substrate, with a mean displacement of $(2D.t)^{1/2}$, when the latter is trapped in a weak-binding ADP.Vi state with vanadate (cf. section 5.3), where D is the diffusion constant and t the elapsed time. These ideas have led to the proposal that high velocities of sliding between motor proteins and their corresponding filaments are achieved via one-dimensional diffusion of weakly attached states. ATP has been proposed to drive a thermal ratchet somewhat akin to the A. F. Huxley 1957 model [46], while slower velocities give time for the crossbridges to enter strongly bound states which enable high isometric tensions to be developed [104].

It is clear from the divergent views currently being advanced that the nature of the crossbridge cycle has yet to be established [see Huxley, 1990; Squire, 1990]. However, technologies are advancing to the state where the behaviour of single crossbridges can be studied, the results of which may well restrict the number of feasible models.

10.2 REGULATION OF MUSCLE CONTRACTION

It follows that detailed understanding of the regulation of crossbridge activity cannot be formulated without a full understanding of the crossbridge cycle itself. Nevertheless, over the last decade there has been reasonable agreement that several regulatory mechanisms function through controlling processes at the myosin active site, rather than simply blocking the binding between actin and myosin. This is most clearly seen in the case of molluscan adductor muscle where the primary events of the mechanism can be monitored in the absence of actin (section 8.3). However, even with actin-linked systems, both biochemical and structural evidence is revealing control points subsequent to myosin head attachment. Some problems remain though. The ATPase activity of isolated S1 is considerably higher than that estimated for an intact, resting vertebrate skeletal muscle based on the low oxygen consumption of the latter (section 4.4). Factors which suppress the basal ATPase activity of the myosin heads in the absence of Ca^{2+} have yet to be identified in this muscle.

Regulation of smooth muscle contraction is also a controversial topic. Myosin light chain phosphorylation is undoubtedly involved in the activation process, but this does not appear to be the whole story. Thin filament regulation has also been characterized [90]. Dephosphorylation of the light chains while the heads are attached has been suggested as the cause of the long-lived latch state in which tension is maintained but the shortening velocity and ATPase activity is reduced by several fold.

The existence, and hence the role, of the 10S conformation *in vivo* has yet to be established. In some smooth muscles there is evidence that activation is accompanied by an increase in myosin filament content, as might occur by the incorporation from a soluble 10S pool. Alternatively, 10S myosin may be concerned with events on a longer timescale associated with growth and development. Even in striated muscles myosin filaments can exchange with monomers on the timescale of several hours to a few days. Fluorescently labelled myosin, injected into developing striated muscle cells, is incorporated into the thick filaments within several minutes and is exchanged out over a period of several hours [105]. Isotope exchange experiments have shown that we have a change of heart every few weeks as a result of new protein synthesis!

As a consequence of the dynamics of the tissue, the study of muscle development has become a major research theme. There is a complex interplay between the physical and chemical stresses placed on a muscle and the expression of contractile proteins. Changing the electrical stimulation pattern of a muscle will cause it to synthesize a new set of contractile proteins [106]. Indeed, failing hearts may be rescued by assistance from fast skeletal muscles which are surgically repositioned and stimulated with a pacemaker. After a period of several weeks, the transplanted muscle shifts its isoforms from those of a fast to a slow type and can rhythmically contract without fatigue [107].

10.3 MOLECULAR GENETIC APPROACHES

The immediate impact of the recent eruption in molecular genetics has been in the rapid determination of nucleotide sequences and hence the derived amino acid sequences of many contractile proteins. Conserved primary structure can often act as a pointer to functional domains in related proteins. The tools of molecular genetics are much more powerful than this, however. Mutations can be exploited as highly selective dissection instruments to cut out one or more functional groups from a protein and examine the consequences. Site-directed mutagenesis has been used to change single amino acid residues in contractile proteins, to delete parts of a gene and to make chimeras of two proteins in an attempt to assign a function to domains within them [see Gerisch *et al.*, 1991; Cross and Kendrick-Jones, 1991]. To exploit these techniques fully requires a reconsideration of the choice of experimental system; rabbits, despite their reputation, reproduce much too slowly and their genome is complex. Of the organisms which contain cross-striated muscles, the fruit fly, *Drosophila*, has the best characterized genome. Effort has therefore been directed towards scaling down mechanical apparatus so that the mechanics of its isolated muscles can be studied [108]. But to exploit genetics also requires a means of mutant selection and here *Drosophila* also has advantages (Fig. 10.1). The actin and myosin which power its flight muscles are specific to these tissues and *Drosophila* does not need to fly in order to reproduce under laboratory conditions. Individuals have therefore been sought with impaired flight ability by relying on natural mutations or by the introduction of specific DNA molecules into the genome through injection into the developing embryo [see Cross and Kendrick-Jones, 1991].

Many mutants with impaired flight are found to have severely disrupted myofibrillar structure and cannot be used to address the nature of the crossbridge cycle. Nevertheless, they may be useful in unravelling the order of assembly of the various myofibrillar elements in the muscle during development. It has been found that thick and thin filaments can assemble independently of each other. Flies with apparently normal myofibrillar structure, but impaired flight, have been isolated by mutagenizing the actin flight muscle gene. Interestingly, a single amino acid change in the large domain of actin, distant from the S1 binding site causes a

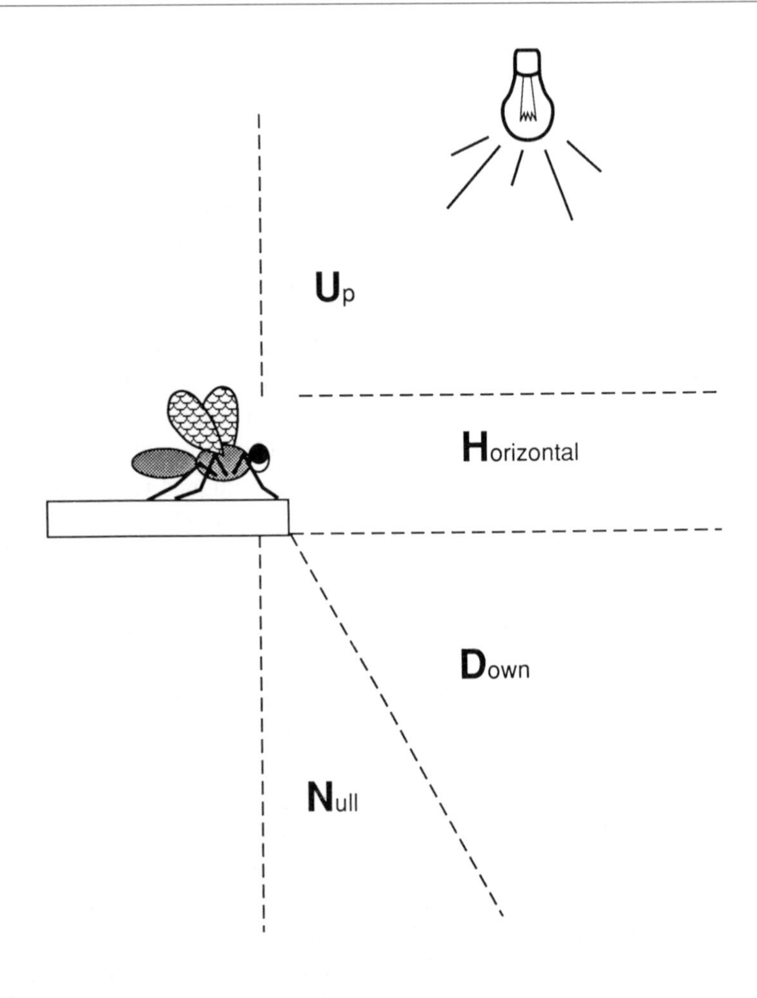

Fig. 10.1 The method of scoring flight ability of wild-type and mutant *Drosophila*. Flies with full flight ability ascend towards the light source while mutants with impaired flight show varying degrees of disability. (Courtesy of Dr J. Sparrow.)

change in the kinetics of the tension generation [108]. This may be cited as evidence for a more direct role of actin in the mechanochemical transduction process than considered previously (cf. section 10.1). However, caution is required in interpretation of any site-directed mutagenesis experiment, since studies of many apparently well-characterized enzymes have produced unexpected changes in catalytic activities when mutagenized far from the active site. Only a small change in conformation is required to alter dramatically the enzymic properties.

With the advent of *in vitro* motility assays, the source of contractile proteins becomes much less restrictive. As a result there are advantages in investigating relatively primitive non-muscle cells in order to exploit the power of their genetic manipulations. Disruption of the gene which codes for myosin II in *Dictyostelium* blocks cytokinesis (cell division) and capping (clustering of receptors), although the cells are still able to divide when grown on a plate because the daughter cells pull themselves apart [see Gerisch *et al*., 1991; Cross and Kendrick-Jones, 1991]. The fact that the cells are still motile, although their chemotaxis response is inefficient, argues that other motor proteins, such as myosin I, are responsible for cell migration in this organism. Having established a myosin II null cell line, it is possible to reintroduce DNA which codes for mutant or truncated myosins by the process of transformation, in order to probe for retrieval of function. These studies have confirmed that a C-terminal region in the tail, which is subject to phosphorylation, is important for disassembly of the myosin filament *in vivo*. Mutant myosins can also be isolated in small amounts and characterized using *in vitro* assays [109].

Expression of the genes for contractile proteins in the bacterium *Escherichia coli* has met with mixed success. Generally, if a protein can be renatured after solubilization in urea, then there is a good chance that it can be recovered in a functional form from an *E. coli* extract. This has been done, for example, with troponin C, myosin light chains and fragments of the myosin rod. However, the myosin head domain and actin have proven more difficult to isolate in a native state, although progress is being made using Sarkosyl to aid renaturation [110]. Once achieved, however, the roles of specific amino acids of proteins from virtually any source are open to study by site-directed mutagenesis. Studies on expressed myosin rod fragments confirm that the C-terminus is vital for filament assembly.

There remains no shortage of problems in the field of muscle contraction. This book can only attempt to scratch the surface of the burgeoning literature.

TOPICS FOR FURTHER READING

Huxley, H.E. (1990) Sliding filaments and molecular motile systems. *Journal of Biological Chemistry,* **265**, 8347–8350. (Includes a discussion of the step-size dilemma.)

Squire, J.M. (ed.) (1990) *Molecular Mechanisms in Muscular Contraction, Topics Molecular and Structural Biology*, Vol 13, Macmillan Press, Basingstoke, London. (Contains chapters covering structural, physiological and biochemical approaches.)

Gerisch, G., Noegel, A.A. and Schleicher, M. (1991) Genetic alteration of proteins in actin-based motility systems. *Annual Review of Physiology*, **53**, 653–681. (A review covering mainly non-muscle systems.)

Cross, R.A. and Kendrick-Jones, J. (eds) (1991) *Motor Proteins. Journal of Cell Science*, Supplement 14. (Includes papers on *in vitro* motility assays and molecular genetic approaches to the study of motility.)

Appendix

Table A.1 Some useful numbers concerning vertebrate skeletal muscle (frog sartorius at 0°C)

Sarcomere length	2.0–3.0 μm (2.5 μm rest length)
Thick filament length	1.6 μm (about 300 myosin molecules)
Thick filament shaft diameter	15 nm
Thin filament length	1.0 μm (about 380 actin molecules)
Thin filament diameter	9 nm
Bare zone length	0.15 μm
Thick filament spacing (centre–centre)	42 nm (at rest length)
Thick–thin filament spacing (″)	22–30 nm (26 nm at rest length)
Thick–thin filament ratio	1:2 (in cross-section)
	1:4 (in total sarcomere)
Total actin content	600 nmol g^{-1} muscle (about 600 μM)
Total myosin content	120 nmol g^{-1} muscle (about 240 μM heads)
ATPase rate (isometric contraction)	0.5 μmol ATP/g muscle/s = 2.1 ATP/myosin head/s
(isotonic contraction)	1.5 μmol ATP/g muscle/s = 6.3 ATP/myosin head/s
(relaxed)	< 0.5 nmol ATP/g muscle/s < 2×10^{-3} ATP/myosin head/s
ΔG for ATP hydrolysis	60 kJ $mol^{-1} = 10^{-19}$ J per molecule
Isometric tension, P_o	20 N cm^{-2} (≅1.6 pN per myosin head)
Elastic modulus (stiffness)	4 kN cm^{-2} (isometric)
	0.025 kN cm^{-2} (relaxed)
Maximum shortening velocity, V_o	2 muscle lengths/s = 2.5 μm/s/half-sarcomere
Maximum power	44 mW g^{-1} muscle (at $V_o/3$)
Maximum thermodynamic efficiency	50% (at $V_o/3$)
Heat production (isometric)	13 mW g^{-1} muscle
Half-time to peak tension	50 ms
Half-time for relaxation	400 ms
Myosin subfragment 1 ATPase	0.01 s^{-1} (in solution)
Actin-activated subfragment 1 ATPase	4.5 s^{-1} (in solution)

Based on references 2, 3, 29, 47 (rates and velocities are about 5 to 10 times higher for rabbit psoas muscle at 20°C)

Table A.2 Some structural proteins of vertebrate skeletal muscle

Protein	*Mol.wt*	*Content (wt %)*	*Location*	*Comments*
Myosin	520 000	43	A-band	motor protein
Actin	42 000	22	I-band	slides over myosin
Tropomyosin	2 × 33 000	5	I-band	regulates actin
Troponin C	18 000	1	I-band	binds Ca^{2+}
Troponin I	21 000	1	I-band	inhibits myosin interaction
Troponin T	31 000	2	I-band	binds to tropomyosin
M protein	165 000	2	M-line	binds to myosin
Myomesin	185 000	< 1	M-line	binds to myosin
Creatine kinase	42 000	< 1	M-line	myosin-bound isoform
C protein	135 000	2	A-band	binds to myosin
F protein	121 000	< 1	A-band	myosin-bound phosphofructokinase
H protein	74 000	< 1	near M-line	binds to myosin
I protein	50 000	< 1	A-band	binds to myosin
α-Actinin	2 × 95 000	2	Z-line	cross-links actin filaments
β-Actinin	37 000	< 1	end I-band	caps actin filaments
Filamin	2 × 240 000	< 1	Z-line	gelates actin filaments
Titin (connectin)	2 500 000	10	A + I bands	links myosin to Z-line
Nebulin	800 000	5	I-band	regulates actin filament length?
Vinculin	120 000	< 1	sarcolemma	peripheral cytoskeleton
Desmin	53 000	< 1	end Z-line	in intermediate filaments
Vimentin	55 000	< 1	end Z-line	in intermediate filaments
Synemin	220 000	< 1	Z-line	associated with desmin
Z-nin	400 000	< 1	Z-line	Z-line cytoskeleton

Based on Amos and Amos (1991) and Ohtsuki *et al.* (1986) (see Chapters 1 and 4 respectively).

References

1 Dawson, M.J., Gadian, D.G. and Wilkie, D. (1977) *J. Physiol.*, **267**, 703–735.
2 Kushmerick M.J. and Davies, R.E. (1969) *Proc. Roy. Soc. Lond., Ser. B*, **174**, 315–353.
3 Kushmerick, M.J. and Paul, R. (1976) *J. Physiol.*, **254**, 693–709.
4 Huxley, A.F. and Niedergerke, R.M. (1954) *Nature*, **173**, 971–973.
5 Huxley, H.E. and Hanson, J. (1954) *Nature*, **173**, 973–976.
6 Peachey, L.D. (1965) *J. Cell Biol.*, **25**, 209–231.
7 Rios, E., Ma, J. and Gonzalez, A. (1991) *J. Muscle Res. Cell Motility*, **12**, 127–135.
8 Fleischer, D. and Inui, M. (1989) *Ann. Rev. Biophys. Biophys Chem.*, **18**, 333–364.
9 Grynkiewicz, G., Poenie, M. and Tsien, R.Y. (1985) *J. Biol. Chem.*, **260**, 3440–3450.
10 Iwazumi, T. (1987) *Am. J. Physiol.*, **252**, C253–C262.
11 Somlyo, A.V., Butler, T.M., Bond, M. and Somlyo, A. P. (1981) *Nature*, **294**, 567–569.
12 Spudich, J.A. (1989) *Cell Regulation*, **1**, 1–11.
13 Somlyo, A.P., Devine, C.E., Somlyo, A.V. and Rice, R.V. (1973) *Phil. Trans. Roy. Soc. Lond. B.*, **265**, 223–229.
14 Vanderkerckhove, J. and Weber, K. (1984) *J. Mol. Biol.*, **179**, 391–413.
15 Tilney, L. (1975) *J. Cell Biol.*, **64**, 289–310.
16 Kabsch, W., Mannherz, H.G., Suck, D., Pai. E.F. and Holmes, K.C. (1990) *Nature*, **347**, 37–44.
17 Holmes, K.C., Popp, D., Gebhard, W. and Kabsch, W. (1990) *Nature*, **347**, 44–49.
18 Pollard, T.D., Doberstein, S.K. and Zot, H.G. (1991) *Ann. Rev. Physiol.* **53**, 653–681.
19 Winkelmann, D.A., Baker, T.S. and Rayment, I. (1991) *J. Cell Biol.*, **114**, 701–713.

20 Walker, M., Knight, P. and Trinick, J. (1985) *J. Mol. Biol.*, **184**, 535–542.
21 Kretsinger, R.H. (1980) *CRC Critical Rev. Biochem.*, **8**, 119–174.
22 Wells, J.A., Sheldon, M. and Yount, R.G. (1980) *J. Biol. Chem.*, **255**, 1598–1602.
23 Elliott, A. and Offer, G. (1978) *J. Mol. Biol.*, **123**, 505–519.
24 Mendelson, R.A., Morales, M.F. and Botts, J.B. (1973) *Biochemistry*, **12**, 2250–2255.
25 Knight, P. and Trinick, J (1984) *J. Mol. Biol.*, **177**, 461–482.
26 Levine, B.A., Moir, A.J.R., Trayer, I.P. and Williams, R.J.P. (1990) in *Molecular Mechanisms in Muscle Contraction*, (ed. J. M. Squire), Macmillan Press, pp. 171–209.
27 Milligan, R.A., Whittaker, M. and Safer, D. (1990) *Nature*, **348**, 217–221.
28 Tokunaga, M., Sutoh, K., Toyoshima, C. and Wakabayashi, T. (1987) *Nature*, **329**, 635–638.
29 Ferenczi, M.A., Homsher, E., Simmons, R.M. and Trentham, D.R. (1978) *Biochem. J.*, **171**, 165–175.
30 Ebashi, S. (1980) *Proc. Roy. Soc. Lond. Ser. B*, **207**, 259–286.
31 Herzberg, O. and James, M.N.G. (1988) *J. Mol. Biol.*, **203**, 761–779.
32 Wang, K. and Wright, J. (1988) *J. Cell Biol.*, **107**, 2199–2212.
33 Engel, P.C. (1981) *Enzyme Kinetics – The Steady-state Approach*, 2nd edn, Chapman and Hall Ltd., London and New York.
34 Gutfreund, H. (1972) *Enzymes – Physical Principles, Wiley-Interscience Ltd, London.*
35 Bagshaw, C.R., Trentham, D.R., Wolcott, R.G. and Boyer, P.D. (1975) *Proc. Nat. Acad. Sci. USA*, **72**, 2592–2596.
36 Trentham, D.R., Eccleston, J.F. and Bagshaw, C.R. (1976) *Q. Rev. Biophys.*, **9**, 217–281.
37 Cremo, C.R., Grammer, J.C. and Yount, R.G. (1989) *J. Biol. Chem.*, **264**, 6608–6611.
38 Houadjeto, M., Barman, T. and Travers, F. (1991) *FEBS Lett.*, **281**, 105–107.
39 Ohno, T. and Kodama, T. (1991) *J. Physiol.*, **441**, 685–702.
40 Sleep, J.A. (1981) *Biochemistry*, **20**, 5043–5051.
41 Mornet, D., Bertrand, R., Pantel, P., Audemard, E. and Kassab, R. (1981) *Nature*, **292**, 301–306.
42 Lymn, R.W. and Taylor, E.W. (1971) *Biochemistry*, **10**, 4617–4624.
43 Stein, L.A., Schwarz, R.P., Chock, P.B. and Eisenberg, E. (1979) *Biochemistry*, **18**, 3895–3909.
44 Hibberd, M.G. and Trentham, D.R. (1986) *Ann. Rev. Biophys. Biophys. Chem.*, **15**, 119–161.
45 Chalovich, J.M., Greene, L.E. and Eisenberg, E. (1983) *Proc. Nat. Acad. Sci. USA*, **80**, 4909–4913.

46 Huxley, A.F. (1957) *Prog. Biophys.*, **7**, 225–318.
47 Huxley, A.F. (1974) *J. Physiol.*, **243**, 1–43.
48 Gordon, A.M., Huxley, A.F. and Julian, F.J. (1966) *J. Physiol.*, **184**, 170–192.
49 Brenner, B. and Eisenberg, E. (1986) *Proc. Nat. Acad. Sci. USA*, **83**, 3542–3546.
50 Reedy, M.K. (1967) *Am. Zoologist*, **7**, 465–481.
51 Huxley, H.E. (1969) *Science*, **164**, 1356–1366.
52 Tsukita, S. and Yano, M. (1985) *Nature*, **317**, 182–184.
53 Trus, B.L., Steven, A.C., McDowell, A.W., Unser, M., Dubechet, J. and Podolsky, R.J. (1989) *Biophys. J.*, **55**, 713–724.
54 Huxley, H.E., Kress, M., Faruqi, A.R. and Simmons, R.M. (1988) *Adv. Exp. Med. Biol.*, (eds H. Sugi and G. Pollack), **226**, 347–352.
55 Maw, M.C. and Rowe, A.J. (1980) *Nature*, **286**, 412–414.
56 Matsubara, I., Yagi, N., Miura, H., Ozeki, M. and Izumi, T. (1984) *Nature*, **312**, 471–473.
57 Fajer, P.G., Fajer, E.A. Schoenberg, M. and Thomas, D.D. (1991) *Biophys. J.*, **60**, 642–649.
58 Cooke, R. (1981) *Nature* **294**, 570–571.
59 Huxley, A.F. and Simmons, R.M. (1971) *Nature*, **233**, 533–538.
60 Huxley, H.E., Simmons, R.M., Faruqi, A.R., Kress, M., Bordas, J. and Koch, M.H.J. (1981) *Proc. Nat. Acad. Sci. USA*, **78**, 2297–2301.
61 Julian, F.J. and Morgan, D.L. (1979) *J. Physiol.*, **293**, 365–378.
62 Goldman, Y.E. and Simmons, R.M. (1977) *J. Physiol.*, **269**, 55–57.
63 Julian, F.J. and Morgan, D.L. (1981) *J. Physiol.*, **319**, 193–203.
64 Higuchi, H. and Goldman, Y.E. (1991) *Nature*, **352**, 352–354.
65 Siemankowski, R.F., Wiseman, M.O. and White, H.D. (1985) *Proc. Nat. Acad. Sci. USA*, **82**, 658–662.
66 Lombardi, V. and Piazzesi, G. (1990) *J. Physiol.*, **431**, 141–171.
67 Eisenberg, E. and Greene, L.E. (1980) *Ann. Rev. Physiol.*, **42**, 293–309.
68 Eisenberg, E. and Hill, T.L. (1985) *Science*, **227**, 999–1006.
69 Brenner, B., Schoenberg, M., Chalovich, J., Greene, L.E. and Eisenberg, E. (1982) *Proc. Nat. Acad. Sci. USA,* **79**, 7288–7291.
70 Pate, E. and Cooke, R. (1989) *J. Muscle Res. Cell Motil.*, **10** 181–196.
71 Webb, M.R., Hibberd, M.G., Goldman, Y.E. and Trentham, D.R. (1986) *J. Biol. Chem.*, **256**, 15557–15564.
72 Goldman, Y.E., Hibberd, M.G. and Trentham, D.R. (1984) *J. Physiol.*, **354**, 577–604.
73 Ferenczi, M.A. (1986) *Biophys. J.*, **50** 471–477.
74 Marston, S.B. and Tregear, R.T. (1972) *Nature*, **235**, 23–24.
75 Marston, S.B., Tregear, R.T., Rodger, C.D. and Clark, M.L. (1979) *J. Mol. Biol.*, **128**, 111–126.

76 Kuhn, H.J. (1981) *J. Muscle Res. Cell Motil.*, **2**, 7–44..
77 Reedy, M.C., Reedy, M.K. and Taylor, K.A. (1990) in *Molecular Mecha nisms in Muscle Contraction*, (ed. J. M. Squire), Macmillan Press, pp. 151–170.
78 Lehman, W. and Szent-Györgyi, A.G. (1975) *J. Gen. Physiol.*, **66**, 1–30.
79 Kress, M., Huxley, H.E., Faruqi, A.R. and Hendrix, J. (1986) *J. Mol. Biol.*, **188**, 325–342.
80 Chalovich, J.M. and Eisenberg, E. (1982) *J. Biol. Chem.*, **257**, 2431–2437.
81 Greene, L.E. and Eisenberg, E. (1982) *Proc. Nat. Acad. Sci. USA*, **77**, 2616–2620.
82 Haselgrove, J.C. and Rodger, C.D. (1980) *J. Muscle Res. Cell Motil.*, **1**, 371–390.
83 Szent-Györgyi, A.G., Szentkiralyi, E.M. and Kendrick-Jones, J. (1973) *J. Mol. Biol.*, **74**, 179–203.
84 Wells, C., Warriner, K.E. and Bagshaw, C.R. (1985) *Biochem. J.*, **321**, 31–38.
85 Vibert, P. and Craig, R. (1985) *J, Cell Biol.*, **101**, 830–837.
86 Wells, C. and Bagshaw, C.R. (1985) *Nature*, **313**, 696–697.
87 Trybus, K.M. (1991) *Curr. Opinion Cell Biol.*, **3**, 105–111
88 Cross, R.A., Jackson, A.P., Citi, S., Kendrick-Jones, J. and Bagshaw, C. R. (1988) *J. Mol. Biol.*, **203**, 173–181.
89 Egelhoff, T.T., Brown, S.S. and Spudich, J.A. (1991) *J. Cell. Biol.*, **112** 677–688.
90 Lehman, W. (1991) *J. Muscle Res. Cell Motil.*, **12**, 221–224.
91 Cooke, R. and Franks, K. (1978) *J. Mol. Biol.*, **120**, 361–373.
92 Kron, S.J. and Spudich, J.A. (1986) *Proc. Nat. Acad. Sci. USA*, **83**, 6272–6276.
93 Warshaw, D.M., Desrosiers, J.M., Work, S.S. and Trybus, K.M. (1990) *J. Cell Biol.*, **111**, 453–463.
94 Toyoshima, Y.Y., Kron, S.J., McNally, E.M., Niebling, K.R., Toyoshima, C. and Spudich, J. A. (1987) *Nature*, **328**, 536–539.
95 Sellers, J.R. and Kachar, B.(1990) *Science*, **249**, 406–408.
96 Kishino, A. and Yanagida, T. (1988) *Nature*, **334**, 74–76.
97 Ishijima, A., Doi, T., Sakurada, K. and Yanagida, T. (1991) *Nature*, **352**, 301–306.
98 Block, S.M., Goldstein, L.S.B. and Schnapp, B.J. (1990) *Nature*, **348**, 348–352.
99 Yanagida, T., Arata, T. and Ohsawa, F. (1985) *Nature*, **316**, 366–369.
100 Harada, Y., Sakurada, K., Aoki, T., Thomas, D.D. and Yanagida, T. (1990) *J. Mol. Biol.*, **216**, 49–68.
101 Uyeda, T.Q.P., Kron, S.J., and Spudich, J.A. (1990) *J. Mol. Biol.*, **214**, 699–710.

102 Homsher, E. (1987) *Ann. Rev. Physiol.*, **49**, 673–690.

103 Berg, O.G. and von Hippel, P.H. (1985) *Ann. Rev. Biophys. Biophys. Chem.*, **14**, 131–160.

104 Vale, R.D. and Ohsawa, F. (1990) *Adv. Biophys.*, **26**, 97–134.

105 Johnson, C.S., McKenna, N.M. and Wang, Y. (1988) *J. Cell Biol.*, **107**, 2213–2221.

106 Brown, W.E., Salmons, S. and Whalen, R.G. (1983) *J. Biol. Chem.*, **258**, 14686–14692.

107 Chiu, R.C. and Bourgeois, I. (eds) (1990) *Transformed Muscle for Cardiac Assist and Repair,* Futura Publishing Co., New York.

108 Drummond, D.R., Peckham, M., Sparrow, J.C. and White, D.C.S. (1990) *Nature*, **348**, 440–442.

109 Manstein, D.J., Ruppel, K.M. and Spudich, J.A. (1989) *Science*, **246**, 656–658.

110 Frankel, S., Condeelis, J. and Leinwand, L. (1990) *J. Biol. Chem.,* **265**, 17980–17987.

111 Offer, G. (1974) in *Companion to Biochemistry*, (eds A.T. Bull, J.R. Lagnado, J. O. Thomas and K. F. Tipton) **1**, 623–671.

Index